Calculus Explorations with *Maple*

to Accompany

CALCULUS

Calculus Explorations with *Maple*

Frank G. Hagin
Jack K. Cohen

to Accompany

CALCULUS

Gerald L. Bradley
Claremont McKenna College

Karl J. Smith
Santa Rosa Junior College

Prentice Hall
Englewood Cliffs, New Jersey 07632

Project Manager: *Joanne E. Jimenez*
Supplements Editor: *Audra Walsh*
Acquisitions Editor: *George Lobell*
Editor in Chief: *Jerome Grant*
Manufacturing Coordinator: *Trudy Pisciotti*

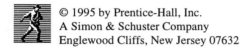

Printed in the United States of America

10 9 8 7 6 5 4 3 2 1

ISBN 0-13-328576-6

PRENTICE-HALL INTERNATIONAL (UK) LIMITED, LONDON
PRENTICE-HALL OF AUSTRALIA PTY. LIMITED, SYDNEY
PRENTICE-HALL CANADA INC. TORONTO
PRENTICE-HALL HISPANOAMERICANA, S.A., MEXICO
PRENTICE-HALL OF INDIA PRIVATE LIMITED, NEW DELHI
PRENTICE-HALL OF JAPAN, INC., TOKYO
SIMON & SCHUSTER ASIA PTE. LTD., SINGAPORE
EDITORA PRENTICE-HALL DO BRASIL, LTDA., RIO DE JANEIRO

Contents

Introduction

REMARKS TO STUDENTS

Maple is an important tool for working scientists. The exact such tool that you will use when you begin your career will no doubt be different, but it makes sense to begin developing skills in using symbolic mathematical computer programs and/or graphing calculators as soon as possible. However, remember that your primary task is to learn calculus, not to become a whiz at symbolic computation. Thus, the projects in this manual focus heavily on a small group of commands and only occasionally "open a door" to more advanced usage. Be aware that *Maple* affords a rich programming environment and that our usage is only a tiny bit of what is available.

We have tried to ease the burden of learning *Maple* by providing online Notebooks containing statements close to those that you will need to complete the project. The idea is to copy and paste these commands into your working Notebook. At the beginning, you will only have to do a small amount of editing on the copied commands to create the ones required by the project. As time goes on, we will shift a bit more of the burden onto you, so that you learn enough of the syntax to function on your own as you use *Maple* in other courses and for your own scientific explorations. If you find that your editing has created problems, it is often easier to recopy the command from the online Notebook than to figure out what *Maple* is complaining about.

However, there is no denying that a complex system like *Maple* will occasionally (we *hope* no more than occasionally!) frustrate you. Please try to keep a sense of humor about the human-computer interface.

This is not a textbook in *Maple*—we have told you only enough to get along. In fact, in some cases, we've written statements and not given an explanation. We assume that you are willing to try things out and observe what they do. For example, observe the difference in *Maple*'s response when we replace the semicolon by a colon in the statement `a := 2;`

```
a := 2;
    2


a := 2:
```

In the projects, we never stopped to explain that the colon prevented *Maple* from echoing the result. We assumed you would find out by trying it. (And now we *have* told you, after all.)

We hope that these projects will add to your calculus course—both by increasing your knowledge and by increasing your interest.

Most projects consist of three parts:

Before the Lab. Relevant reading and warm-up problems. These problems are important to the course as well as to the project.

In the Lab. Problems requiring *Maple*. Your instructor will tell you which problems to do. This section should take about an hour of lab work.

After the Lab. You'll be asked to reflect on what your computations revealed.

This Project Booklet includes:

1. An introductory tutorial on the basics of using *Maple* in a Notebook environment. Our version is explicitly written for the NeXT environment—other *Maple* Notebook environments are similar.

2. Weekly projects covering a three semester calculus course—there are "extra" projects to allow for differing taste and for variety of choice between semesters or sections.

3. Summaries of *Maple* statements.

4. A diskette containing a *Maple* "notebook" for each project containing *Maple* statements or code serving as copy and paste templates.

Your instructor may well shorten some of the projects by telling you to omit certain exercises. We suggest that you read over the omitted exercises and give brief thought as to how you would tackle them. You will also note that we have labeled certain projects as "optional." This designation is given to projects that either use advanced *Maple* commands that won't be needed in the sequel, or that involve a greater degree of exploration on your part. Again, even if your instructor omits some or all of these projects because of time pressure, we suggest that you read them—you may well want to refer back to these projects later in your academic or scientific career.

Your Authors welcome your comments for improving this manual. Our Internet addresses are:
jkc@keller.mines.colorado.edu
fhagin@newton.mines.colorado.edu

REMARKS TO INSTRUCTORS

For the past three years at the Colorado School of Mines, all 500-plus freshmen have taken a calculus course using major portions of the material you have in your hands. Your authors each have over 25 years of teaching experience at all university levels from freshman to Ph. D. level graduate courses. We have participated fully in all aspects of university life and are well published applied mathematicians. However, our recent experience with this calculus program ranks among the most exciting and rewarding of our careers—it added a distinct element of fun compared with our previous calculus experiences. However, while "fun" is not to be underestimated as a motivating force, we have also kept firmly in mind a number of more concrete pedagogical objectives. We explain our viewpoint in the following paragraphs.

Our guiding star is that we are teaching calculus, not *Maple*. This has important implications. For example, to our mind, premature reliance on constructs such as `limit` and `fsolve` can easily degenerate the *Maple* experience to being a high tech analog of looking up the answer in the back of the book. These "black boxes" don't help teach calculus, so we relegate them to a secondary role. On the other hand, we believe that students should be aware of these tools for use *after* they master the concepts of calculus, so we do include some projects devoted to such commands. The instructor notes to such projects will clearly state these projects are not of primary importance to learning calculus and may be skipped. In particular, succeeding projects will *not* depend on covering these "black box" projects. Consider at least making some reading assignments in some of these projects, but don't let them deflect you from your main task of teaching calculus.

To minimize the numerous syntax errors that occur when the neophyte tries to type in unfamiliar statements, we provide on-line template commands that allow the student to copy, paste and then do only minor editing to obtain the needed adaptation of the command. While we do not emphasize *Maple* syntax, over time we gradually make the students sensitive to it by gradually decreasing the amount of template help for the core *Maple* commands that have already occurred several times earlier. Occasionally, we use *Maple* commands outside the minimal core to "open doors" for students—but these more advanced commands are always supported fully by a template.

We have firm objectives in teaching calculus and these projects are designed to reinforce them:

Solve general problems. Instead of solving problems with numbers that make the answers "come out nice," solve families of problems (i.e. problems with parameters).

Emphasize fundamentals. Don't rush through hundreds of topics, instead take time with the topics that deserve it, and train students to *use* their text, when necessary, to look up topics that are less important.

Empower students. The picture of two or three students huddled around a computer (cooperative education) is often a better picture than 35 students busily taking notes in a lecture hall. We allow our students to hand in joint work for all the *Maple* projects—and we encourage you to think about doing this too.

As stated above, most projects consist of three parts: **Before the Lab**, **In the Lab** and **After the Lab**. The intent of this division is to make the Laboratory experience meaningful.

For each project, the instructor's Answer Booklet contains:

a) Answers (naturally!).

b) An introductory statement of the goals of the project.

c) Suggestions for ways to shorten the project (you'll probably want to vary the assignments from semester to semester or section to section).

d) Occasional Instructor Notes about likely student questions, etc.

The Answer Booklet is accompanied by an instructor diskette containing the commands needed to reproduce the results cited in the answers and a set of *Maple* utilities for constructing classroom demonstrations.

We have been generous in both the number of projects and their contents. For most projects, we list exercises that can be omitted to yield a shorter project. We have also labeled certain projects as "optional." This designation is given to projects that either use "black box" *Maple* commands that won't be needed in the sequel or that introduce modest "discovery" explorations. These projects can be among the most exciting, so we hope you won't omit *all* the optional projects. Similarly, we do not expect that you will assign *all* the regular projects. Also, notice that the generosity in projects is greater earlier in the Manual—this is to accommodate courses with differing priorities. Consequently, you may wish to consider assigning an "optional" project that appears early in the Manual in (say) the third semester, with a view towards increasing competence in *Maple* at that later time.

ACKNOWLEDGMENTS

Over these past three years, we have written numerous worksheets and projects containing problems to be solved using *Maple*. Most of them have significant original content, but many have been motivated by or use material from articles written in teaching-oriented journals. In particular, many of our project questions originated from routine textbook exercises. Similar exercises could be found in many calculus texts, but since we used the popular Edwards and Penney Third Edition during the past three years, often the original routine kernel exercise was drawn from this excellent text and we wish to acknowledge that.

We owe a great debt to our colleagues on the "Calculus Team" at the Colorado School of Mines, who willingly tested earlier versions of these projects. In particular, the advice and encouragement of the team coordinator, Barbara Bath, was invaluable. We also wish to acknowledge our grader, Shelby Worley, for his astute comments about student difficulties with the earlier versions of these projects. Our questions are distinctly more "student friendly" because of his efforts. Finally, we give our thanks to the students of the School of Mines for their enthusiasm and advice.

ADDRESSES

We welcome your comments for improving this manual.

Jack K. Cohen   Frank G. Hagin
Center for Wave Phenomena   Department of Mathematics
Colorado School of Mines   Colorado School of Mines
Golden, CO 80401   Golden, CO 80401
jkc@keller.mines.colorado.edu   fhagin@newton.mines.colorado.edu

Calculus Explorations with *Maple*

to Accompany

CALCULUS

Tutorial Exercises

BEFORE THE LAB

Be sure to carefully read the online file "Tutorial.ms" before attempting this project. **Tip:** The on-line notebooks that accompany these projects have "templates" for you to "copy and paste" from. This avoids little errors like using round brackets when you should use square brackets. Eventually, you'll have to write expressions on your own, but, for now, we'll give lots of help to minimize the inevitable frustration of learning a new language (*Maple*).

IN THE LAB

Exercise 1 Use *Maple* to evaluate $\sin(\pi/3)$ *numerically*. Verify (somehow) that the answer is correct.

Exercise 2 Evaluate $\sqrt{2}$ numerically to 16 "significant figures" (i.e, use 16 as the optional second argument in the numerical command, `evalf`).

Exercise 3 Define the functions $f(x) = x^2 + 1$ and $g(x) = x^3 - x^2 - 9x + 9$ and plot them on the same graph. From your graph estimate the three values of x where the plots cross.

Exercise 4 Form the "rational" function, $r(x) = g(x)/f(x)$ and plot it for $-20 \le x \le 20$. Make a hand sketch of what you see.

Exercise 5 Define the function $G(x) = x^3 + 2x^2 - 15x - 30$ and plot G on several x-intervals to get a good idea of its behavior.

  a) Zoom in on the largest x-value for which $G = 0$. Does the function become "almost linear" as you zoom in?

  b) Use the `factor` command to find the exact x-values at which $G = 0$.

  c) Define a new rational function, $R = G(x)/F(x)$, where $F(x) = x^4 + 3$. Decide if R is "almost linear" for x large (as was the case with the r discussed earlier).

AFTER THE LAB

Exercise 6 Explain why the graphs of $g(x)$ and $r(x)$ are similar for "small" x.

Exercise 7 The graph of $r(x)$ looks linear for "large" x. Can you figure out *what* linear function approximates r for large x?

BEFORE THE LAB

In high school, you learned the solution formula for the quadratic equation, $Ax^2 + Bx + C = 0$. However, you may not know any methods for solving higher degree or transcendental equations (i.e., equations involving trig functions, logs, etc.). Oftentimes, such equations can be approximately solved by using `plot` to "zoom in" on the desired root as you will do in the laboratory portion of this Project.

Exercise 1 In each case, write an equation of the straight line L that is described:

   a) L passes through $(1, 5)$ and is parallel to the line with equation $2x + y = 10$.

   b) L passes through $(-2, 4)$ and is perpendicular to the line with equation $x + 2y = 17$.

Remark: You may find it useful to check your text for the various forms used to describe straight lines: slope-intercept, point-slope, and two-point, to find the easiest one to use for the above problems.

Exercise 2 You could make up a million problems like the ones in the previous exercise. It makes more scientific sense to solve *general* problems (though "warm-ups" with numerical values are often useful). Solve the following generalized version of the first part of Exercise 1:

   Write the equation of the straight line L that passes through the point (p, q)
   and is parallel to the line with equation $ax + by = c$. Check that your answer
   "works" with the specific numbers in part (a) of Exercise 1.

Exercise 3 State the general problem corresponding to the second part of Exercise 1, solve it, and check with the specific example.

We will soon learn that:

> The slope of the tangent line to the parabola $y = x^2$ at $x = a$ is $2a$.

Thus, the slope of this parabola at $x = 1$ is 2, the slope at $x = 3$ is 6 and so on. Please accept this as a fact for now.

Illustrative Exercise: Find the equation of the line through the point $P(2, 0)$ that is normal to the parabola $y = x^2$.

   Solution: Draw a figure (see Figure 1) that includes a rough sketch of the
   normal from the given point. Introduce the letter a to denote the x-coordinate
   of the point where the normal cuts the parabola. Thus, the point (a, a^2) is on
   both the parabola and the normal. The slope of the parabola at $x = a$ is $2a$, so
   the slope of the normal is the negative reciprocal value $-1/2a$. Equating this
   latter slope to the slope from (a, a^2) to the given point $(2, 0)$ yields

$$-\frac{1}{2a} = \frac{a^2 - 0}{a - 2}$$

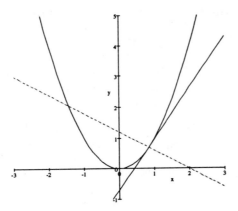

FIG. 1. Sketch for the Illustrative Exercise.

leading to the *cubic* equation,

$$2a^3 + a - 2 = 0$$

for a. We can see from our sketch that $a \approx 1$, but this is *not* the exact root (try it). Instead of resorting to trial and error, we can `plot` the function near $a = 1$ and then "zoom in" on the root. The three successive `plot` commands

```
plot(2*a^3 + a - 2, a = 0.7..1.1);
plot(2*a^3 + a - 2, a = 0.82..0.85);
plot(2*a^3 + a - 2, a = 0.834..0.836);
```

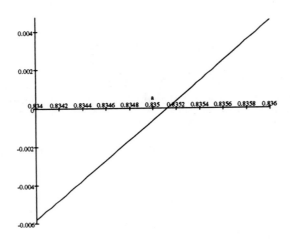

FIG. 2. Zooming in on the solution of the Illustrative Exercise.

produce the results shown respectively from top to bottom in Figure 2. From the bottom graph, we see that $a \approx .835$. Now is the time to be careful— we've worked hard to get a good approximation to a, but that is *not* quite what the exercise called for! The correct solution to the problem is to produce the requested *normal* from $(2,0)$ to the parabola. This line is given by $\boxed{y = -\frac{1}{2a}(x-2), \quad \text{where } a \approx .835}$.

Exercise 4 One problem solving maxim we've ignored in the above solution is that of trying to avoid the use of specific numbers. To maintain generality, let (p,q) denote the given point (i.e replace $(2,0)$ by (p,q)). Show that the cubic equation giving the point $x = a$ where the normal from (p,q) intersects the parabola $y = x^2$ is $\boxed{2a^3 - (2q-1)a - p = 0}$. It's good practice to check that this agrees with the specific case worked out in the Illustrative Exercise, so do it!

IN THE LAB

Exercise 5 In reference to the previous exercises, find the normal when $(p,q) = (4,0)$. If you are uncertain how to proceed or how to present your solution, re-read the Illustrative Exercise.

Exercise 6 In reference to the previous problems, when $(p,q) = (1, 17/4)$ show that there are *three* normals and determine them by doing three "zooms." Notice that the given point is *inside* the parabola.

Notice how much more convenient and less error prone it is to use the *general* **form for the cubic instead of setting up and simplifying the fractions for each specific point** $(4,0)$**,** $(1, 17/4)$**, ...**

Exercise 7 Show that the cubic equation in Exercise 6 has the *exact* solution $a = 2$. Use `Factor` on your cubic to obtain a quadratic equation. Solve the quadratic to get all three roots of the cubic exactly. Check that the exact solutions agree with the ones you found by the zoom method.

AFTER THE LAB

Exercise 8 In light of your experience with the previous exercises, comment on the efficiency of the zoom method if highly accurate solutions (say 10 decimal place accuracy) are required.

Exercise 9 Comment on the efficiency of the zoom method if solutions are required for *many* different points (p,q).

Exercise 10 In light of the results obtained in the lab exercises, a natural conjecture is that there are three normals for points interior to the parabola $y = x^2$ and only one for exterior points. Investigate the truth of this conjecture for the special case when the point is on the y-axis (that is, $p = 0$).

Exercise 11 As a review of hand methods, repeat Exercise 7 using the algebraic analog of long division to divide the known factor out of the cubic and thus again obtain the quadratic equation for the remaining roots.

The Asteroid Problem

BEFORE THE LAB

Exercise 1 You land in your space ship on a spherical asteroid. Your partner walks 400 meters away along the smooth surface carrying a one meter rod, and thereby vanishes over the horizon. When she places one end on the ground and holds the rod straight up and down, you, lying on your stomach, can just barely see the tip of the rod. The ultimate quest is to determine the radius r of the asteroid, but, for now, just draw a careful diagram and label it with the quantities you think will be useful in getting an equation involving only r and known quantities.

Exercise 2 After checking with your instructor that your diagram for the asteroid problem is correct, use it to derive the equation for the radius r in terms of the given quantities. Again use symbols instead of specific numbers and check your equation with your instructor before going on with the project.

IN THE LAB

Building a table in *Maple*

There are several ways to build a table. Here we use the `for` command:

```
for x from 0 by 0.2 to 1.0 do
    lprint(x, cos(x) -x)
od;
 0    1
.2    .7800665778
.4    .5210609940
.6    .2253356149
.8   -.1032932907
1.0  -.4596976941
```

Recall one could also use, in the above `for` statement, the set notation {x, cos(x)-x}, or print(x, cos(x)-x).

Exercise 3 Give an intuitive explanation of why there is a solution of $\cos x = x$ in the interval $0.6 \leq x \leq 0.8$; and use a table to determine this solution with error less than 0.05.

The Asteroid Problem—continued

Exercise 4 You now know the correct equation for r. Using **plot** and/or tables, find the solution to one decimal place accuracy for the given numerical values.

Comment: How much would you contribute to the Indigent Math Professors' Support Fund for an easily derived estimate of the solution to within 1 meter? Actually, this fee is covered in your tuition for this course—stay tuned. Without questioning that **plot** and tabling are terrific tools, your experience with the current problem may cause you to wonder if there aren't some better methods of solving equations. If so, you will be glad to hear that you will, indeed, be learning about more advanced solution methods this semester.

AFTER THE LAB

We will be returning to the asteroid problem several times later in the semester, so it is worthwhile to reflect on your experience so far:

Exercise 5 In light of your Laboratory investigations, discuss how you could have derived the answer more efficiently.

Exercise 6 In light of your Laboratory investigations, discuss the virtues and limitations of the `plot` command and tabling in solving equations.

BEFORE THE LAB

Your text shows that the derivative of $f(x) = x^2$ is $f'(x) = 2x$. This is a special case of the **Power Law**:

$$f(x) = x^n \implies f'(x) = nx^{n-1} \qquad \text{for } n = 1, 2, \ldots.$$

The `for` loop

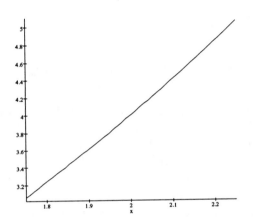

FIG. 3. First four zooms of x^2 at $x = 2$.

It is tiring to generate a series of "zoom" `plots` one at a time. Here is a `for` loop that zooms in on $x = 2$ for $f(x) = x^2$:

```
a   := 2.0: h := 4.0:
for k to 5 do
    h := h/2:
    plot(x^2, x = a-h..a+h)
od;
```

On each pass through the loop, a `plot` is generated with a plot range varying from $x = a - h$ to $x = a + h$. The quantity h is set to 4 before the loop. As we enter the loop, it is cut in half, so the first plot goes from $x = a - 2$ to $x = a + 2$. Before the second plot, h is cut in half again, so that the second plot goes from $x = a - 1$ to $x = a + 1$. Similarly, the third plot goes from $x = a - 1/2$ to $x = a + 1/2$ and so on. Since a is set to 2 before the loop, the actual successive numerical plot ranges are $[0, 4]$, $[1, 3]$, $[1.5, 2.5]$, $\ldots$. So we are, indeed, "zooming" in on $x = 2$.

The above `for` loop is more concise than the usual "`for k from 1 by 1 to 5 do ...`"; since we are starting with $k = 1$ and taking steps of 1, the 'from' and 'by' parts can be omitted. (That is, 1 is the default value for the 'from' and 'by' clauses). Since we are to pass through the loop 5 times the final plot range is $[1.875, 2.125]$. The `plots` from the first four zooms are shown in Figure 3 and the final zoom is shown in Figure 4.

Tip: It is sensible to set the "iterator" limit (here, 5) to a small number (like 2) until you are *sure* the loop is doing what you want.

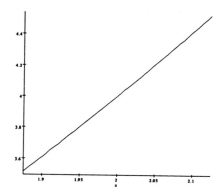

FIG. 4. Final zoom of x^2 at $x = 2$.

Exercise 1 Use the output of the final step in the **for** loop above shown in Figure 4 to verify that the slope of $f(x) = x^2$ at $x = 2$ is what it should be. Watch out for the fact that *Maple* rescales the **plot**—you should use the *numbers* on the axes for your computations, *not* the apparent slope. Reproduce the figure as a rough sketch and draw any necessary lines on your sketch. Make the best estimate you can of the slope and also state the error in your approximation to the exact slope. **Warning**: Take into account the fact that *Maple* does not always put the origin where the axes cross—that is, be sure to compute the difference quotient, not just y/x.

IN THE LAB

Exercise 2 Use *Maple* to verify the power law for $n = 3$ and $n = 4$. Method: Define a suitable **f(x)** and simulate the standard simplification of the difference quotient by using the command **expand((f(x + h) - f(x))/h)**. Then evaluate the resulting simple limit as $h \to 0$ any way you like.

Exercise 3 In this exercise, we find the tangent line to a curve using the idea that the tangent line is the *limit* of secant lines. As a nontrivial example, we use the function:

$$f(x) = \frac{(x^3 - 5)(x^2 - 1)}{x^2 + 1}$$

and find the tangent line at the specific point $(2, f(2)) = (2, 1.8)$.

To this end, consider two points on the graph, namely $(2, f(2))$ and $(b, f(b))$, where b is close to, but not equal to 2. The slope of the "secant" line (the line that connects the two points) is $(f(b) - f(2))/(b - 2)$. Let's start with $b = 2.5$. To calculate the secant line, you execute the following commands:

```
f := x -> (x^3 - 5)*(x^2 - 1) / (x^2 + 1);
b := 2.1:
slope := (f(b) - f(2)) / (b - 2.0):
secant := f(2.0) + slope*(x - 2.0);
```

To simultaneously display the graph of f and its secant line near the point of interest, execute:

```
plot( {secant, f(x)}, x = 1.5..3.0 );
```

You are to proceed, taking b closer and closer to 2.0 until you have what looks like a tangent to the curve at $(2, f(2))$—that is, find a value of b (close to, but not equal to 2.0) such that you effectively have a tangent. Write down the values of b and $f(b)$, and turn in the plot of the curve and its "tangent" you are most satisfied with.

Exercise 4 Use the approach of Exercise 1 to approximately determine the slope and tangent line to each of $f(x) = x^3$ and $f(x) = x^4$ at the point $x = 2$. It is probably easiest to print the *one* plot for each function that will allow a good estimate of the slope and later make any necessary measurements with a straight-edge.

Exercise 5 Use the method of the previous exercise to again solve Exercise 3, that is, find the tangent line to $f(x) = (x^3 - 5)(x^2 - 1)/(x^2 + 1)$ at $x = 2$ by zooming in on $x = 2$.

Exercise 6 Use the method of the previous two exercises to explore the slope and tangent line to

$$f(x) = (x - 2)^{2/3} + 2x^3$$

at $x = 2$. You will need more zooms than previously to unearth the behavior near $x = 2$—in particular, what does the slope appear to be after 4, 8 and 12 zooms? Continue zooming until you are sure you've captured the true microscopic behavior near $x = 2$. Is there a unique tangent line at $x = 2$? Unique slope at $x = 2$? Why do you think so many zooms were needed?

Tip: In *Maple*, fractional exponents must be written with parentheses, for example, `(x - 2)^(2/3)`. Omitting the parentheses changes this to $\frac{1}{3}(x - 2)^2$.

Exercise 7 Find the (approximate) slopes of $f(x) = \sin x$ at the points $x = \pi/3$ and $x = \pi/6$.

Tip: In *Maple*, $\sin x$ is written `sin(x)` and π is written `Pi`.

AFTER THE LAB

Exercise 8 Use the power rule (as cited above) to check that the numerical values you found for the slopes of x^3 and x^4 were approximately correct.

Exercise 9 We haven't yet discovered the formula for the derivative of $\sin x$, but look it up in your textbook and check the results you got in Exercise 7.

BEFORE THE LAB

Exercise 1 Consider the following piecewise defined function:

$$f(x) = \begin{cases} x^3 + 2 & \text{if } x \leq -1; \\ x^4 + 1 & \text{if } x \geq 1; \\ x^2 + x + 1 & \text{elsewhere.} \end{cases}$$

a) Find out where $f(x)$ is continuous and discontinuous and justify your conclusions.

b) Determine values b and c such that the related function

$$f(x) = \begin{cases} x^3 + 2 & \text{if } x \leq -1; \\ x^4 + 1 & \text{if } x \geq 1; \\ x^2 + bx + c & \text{elsewhere.} \end{cases}$$

is continuous everywhere.

IN THE LAB

Exercise 2 Hand in the output from these *Maple* commands:

```
f :=   x -> x^3 + 3*x^2;
f(2);
subs(x=2, f(x));
f(t);
f(cat);
for n to 5 do
   print(f(n))
od;
```

Exercise 3 Use `factor` to compute the following limits:

a) $\displaystyle \lim_{x \to 13} \frac{x^3 - 9x^2 - 45x - 91}{x - 13}$

b) $\displaystyle \lim_{x \to 13} \frac{x^3 - 9x^2 - 39x - 86}{x - 13}$

c) $\displaystyle \lim_{x \to 13} \frac{x^4 - 26x^3 + 178x^2 - 234x + 1521}{x - 13}$

Exercise 4 Find

$$\lim_{x \to 0} \frac{\sqrt{25 + 3x} - \sqrt{25 - 2x}}{x}$$

by each of the following methods:

a) Make a `plot` of the fraction near the point $x = 0$. This often "works" even when, as in this case, the function is not defined at the critical point. You may generate a complaint if `plot` happens to try the exact value $x = 0$—but nonetheless, the `plot` will be fine.

b) Make a table near $x = 0$.

c) The straightforward use of a table doesn't "home in" on the critical point. Here is some advanced *Maple* code that is more tuned for this purpose. Note we are approaching the limit value a by increasing powers of $1/2$.

```
f := x -> (sqrt(25 + 3*x) - sqrt(25 - 2*x))/x;
a := 0: ntimes := 5:
for k to ntimes do
    print( evalf(a + 1/2^k), '    ', f(evalf(a + 1/2^k)) )
od;
```

Try this code out. As it stands, the code explores the *right* limit; modify it to explore the left limit and report the results of running your modified code.

Exercise 5 Evaluate the following limits as $x \to 0$ using any one of the methods introduced so far:

a) $\dfrac{x^3 - x^2 - 4x + 4}{x - 1}$

b) $\dfrac{\sin x}{x}$ (This is an important limit.)

c) $\dfrac{1 - \cos x}{x}$ (So is this.)

d) $\dfrac{\sin 5x}{x}$

e) $(1 + x)^{1/x}$ (Another important limit.)

Exercise 6 Verify the assertions you made in Exercise 1 by handing in a `plot` for each part of the problem. Here is code for f in the first part:

```
f := proc(x)
  if x <= -1 then x^3 + 2
  elif x >= 1 then x^4 + 1
  else x^2 + x + 1
  fi;
end;
```

Do something similar for the second part.

AFTER THE LAB

Exercise 7 Algebraic limits of the 0/0 form can be simplified to a form where the limit can be obtained by just evaluating the function. As we have seen, sometimes standard tools such as factoring can be used advantageously. Use the trick of multiplying top and bottom by a sum of square roots ("rationalizing the numerator") to algebraically check the result of Exercise 4 and check that your Lab results were correct.

Exercise 8 If, as claimed in the previous exercise, "algebraic limits of the 0/0 form can be simplified to a form where the limit can be obtained by just evaluating the function," what about

$$\lim_{x \to 8} \frac{x^{2/3} - 4}{x - 8} ?$$

Hint: Factor (i.e., divide) the quantity $x^{1/3} - 2$ from both numerator and denominator.

Exercise 9 You have noticed that many limits can be obtained by just evaluating the given function at the given point. State precisely a condition under which this is valid. Which of the following limits can be evaluated in this simple way? Justify your answers.

a) $\lim_{x \to 5} \dfrac{x + 5}{x^4 + x^2 + 1}$

b) $\lim_{x \to 5} \dfrac{x^2 - 25}{x - 5}$

c) $\lim_{x \to 5} \sqrt{x^6 - 5}$

Exercise 10 The function in Exercise 1 seems artificial. Can you think of any cases in science where a function would naturally be defined piecewise?

Limits With Maple (optional)

BEFORE THE LAB

Exercise 1 Explain why the following limits are equivalent:

1. $\lim\limits_{x \to 0+} f(x)$

2. $\lim\limits_{u \to +\infty} f(1/u)$

Explain (intuitively) why the latter limit implies the value of the limit $\lim_{n \to \infty} f(1/n)$, where now n is considered to be an integer.

IN THE LAB

Exercise 2 Use both the `limit` command and the special table code to evaluate

a) $\lim\limits_{x \to 0} \left(\dfrac{1}{\sin^2 x} - \dfrac{1}{x^2} \right)$

b) $\lim\limits_{x \to 0} \left(\dfrac{\sin(\tan x) - \tan(\sin x)}{x^7} \right)$

Explain why the results by the two methods are, or are not, consistent.

Exercise 3 Use the `limit` command to evaluate the limits of the following functions as $x \to 0$:

a) $\dfrac{x^3 - x^2 - 4x + 4}{x - 1}$

b) $\dfrac{\sin x}{x}$

c) $\dfrac{1 - \cos x}{x}$

d) $\dfrac{\sin 5x}{x}$

e) $(1 + x)^{1/x}$

If you did Exercise 3 in the *Limits and Continuity* project, comment on the consistency with your previous results, otherwise check consistency using the special table command. **Caution**: Fractional exponents in *Maple* need parentheses. Write (1 + x)^(1/x), *not* (1 + x)^1/x. The latter translates to $(1 + x)/x$—quite a different animal.

Exercise 4 Use `limit` to evaluate the limits of the following functions as $x \to 0$:

   a) $(1 + 2x)^{1/x}$

   b) $(1 + 3x)^{1/x}$

   c) $(1 - 2x)^{1/x}$

Draw a general conclusion and try to verify it.

Exercise 5 Use `limit` to evaluate the limits of the following functions as $x \to 0$:

   a) $(1 + x^2)^{1/x}$

   b) $(1 + 2x^2)^{1/x}$

   c) $(1 + x^3)^{1/x}$

   d) $(1 + x^{3/2})^{1/x}$

   e) $(1 + \sqrt{x})^{1/x}$

Draw a general conclusion.

Exercise 6 Use `limit` to evaluate the limits of the following functions as $x \to 0$:

   a) $\dfrac{\sin x}{x}$

   b) $\dfrac{\sin 2x}{x}$

   c) $\dfrac{\sin 3x}{x}$

Draw a general conclusion and try to verify it.

AFTER THE LAB

Exercise 7 Rewrite the limits in Exercise 4 as limits involving $n \to \infty$. Similarly, re-express your general conclusion.

Exercise 8 Using your previous work, evaluate:

   a) $\lim\limits_{h \to 0} (1 + hx)^{1/h}$

   b) $\lim\limits_{n \to \infty} \left(1 + \dfrac{x}{n}\right)^n$

BEFORE THE LAB

The Binomial Theorem is:

$$(a+b)^n = a^n + \frac{n}{1}a^{n-1}b + \frac{n \cdot (n-1)}{1 \cdot 2}a^{n-2}b^2 + \ldots + b^n. \tag{1}$$

Exercise 1 There are "dots" in the above expression for the binomial theorem. To be sure that you understand the correct pattern, write the first missing term explicitly (i.e. the one with the b^3 in it). Also write the last missing term explicitly (i.e. the one with the $a^1 = a$ in it).

Exercise 2 Show that your answers to the previous exercise are correct by expanding out $(a+b)^n$ for $n = 1, 2, 3$ and 4.

Exercise 3 Write the binomial theorem for the special case when $a = 1$ and $b = x$. This is the form we'll use below.

Exercise 4 By factoring out a^n, show that (with a slightly different definition of x), the general binomial theorem in Equation 1 can be written in the form of the previous exercise (so it isn't so special after all).

Exercise 5 Some of the founders of calculus were intrigued by the possibility of using the binomial theorem for *fractional* exponents. Considering only "small" x, try to discover what value of the parameter a is appropriate in

$$\sqrt{1+x} \approx 1 + ax?$$

Hint: Square both sides. Does your value of a agree with the binomial theorem for $n = 1/2$? Assuming that the binomial theorem for integer n is a reliable guide for fractional n too, write the next term of the approximation.

Exercise 6 For integer n, the binomial theorem has a finite number of terms (exactly $n+1$ terms, in fact). Assuming that the theorem makes sense for fractional n, do you think that there is a hard and fast last term? Explain.

More on the `for` Loop

In the *Derivatives, Slopes and Tangent Lines* project, we used a `for` loop with a simple numerical iterator indicating how many times the loop is to be executed. More typically, the `for` loop uses a "counter" (here k):

```
sum := 0.0;
for k from 1 by 2 to 7 do
   sum := sum + k^2
od;
```

This loop computes $1^2 + 3^2 + 5^2 + 7^2$ (that is, k starts at 1 and is incremented by 2 repeatedly until it gets up to 7).

IN THE LAB

Exercise 7 From your "Before the Lab" work, you know that using only two terms of the binomial theorem gives $(1+x)^5 \approx 1+5x$. Is this really a usable approximation? One way to start exploring this question is to **plot** the two functions $f(x) = (1+x)^5$ and $g(x) = 1+5x$ on the same graph for the three intervals $0 \le x \le 0.1$, $0 \le x \le 1.0$, and $0 \le x \le 10.0$. What about *negative x*?

Exercise 8 Investigate the accuracy of $\sqrt[3]{1+x} \approx 1 + \frac{1}{3}x$ for small x along the lines of Exercise 7.

Exercise 9 Investigate the "infinite series,"

$$S = 1 + \frac{1}{2^1} + \frac{1}{2^2} + \frac{1}{2^3} + \dots .$$

Try to approximate S. You might use code like the following. Note the 'printlevel := 0' command is used to suppress the printing of the results of the 'S := ...' command each time through the loop. It is a good idea to set this back to the value 1 when done with this construct.

```
k := 'k': S := 'S':
lprint('k', '          ', 'S');
S := 1:
printlevel := 0:
for k to 20 do
   S := S + 1.0/2^k:
   lprint( k, '          ', evalf(S,4) )
od;
```

Exercise 10 Similarly investigate the series,

$$S = 1 + \frac{1}{2^2} + \frac{1}{3^2} + \frac{1}{4^2} + \dots .$$

Caution: For this problem, the previous code needs to be changed in *two* places.

Exercise 11 Similarly investigate the series,

$$S = 1 + \frac{1}{2} + \frac{1}{3} + \frac{1}{4} + \dots .$$

AFTER THE LAB

Exercise 12 Consider the infinite series,

$$S = 1 - 1 + 1 - 1 + 1 - 1 + \dots .$$

If we group like this:

$$S = (1 - 1) + (1 - 1) + (1 - 1) + \dots ,$$

then we get $S = 0$. On the other hand, if we group it like this:

$$S = 1 + (-1 + 1) + (-1 + 1) + (-1 + 1) + \ldots,$$

we get $S = 1$. We seem to have created something (1) out of nothing (0), what do you think about this?

Exercise 13 It seems reasonable to ask if the binomial theorem works for *negative* exponents. Investigate the case $n = -1$. Have you seen this somewhere before?

Exercise 14 Evaluate the limit

$$\lim_{x \to 0} \frac{\sqrt{25 + 3x} - \sqrt{25 - 2x}}{x}$$

from the *Limit and Continuity* project by using the idea described in Exercises 4 and 5.

BEFORE THE LAB

The derivative command `diff` has a standard *Maple* syntax:

```
diff(x^3, x);
```

$$3 \, x^2$$

Alternatively, you can define an expression beforehand:

```
y := 1/(1 + sqrt(1 + sqrt(x))):
diff(y, x);
```

$$- \frac{1}{4 \, (1 + (1 + x^{1/2})^{1/2})^2 \, (1 + x^{1/2})^{1/2} \, x^{1/2}}$$

or define a full-fledged function:

```
f := x -> (1 + (1 + x)^2)^3:
diff(f(x), x);
```

$$3 \, (1 + (x + 1)^2)^2 \, (2 \, x + 2)$$

In the latter case, you can use instead the D operator; but it requires some care. For example, D(f) produces a *function*, not an *expression*. To get an expression you need D(f)(x).

```
D(f);
```

$$x \rightarrow 3 \, (1 + (1 + x)^2)^2 \, (2 + 2 \, x)$$

```
D(f)(x);
```

$$3 \, (1 + (x + 1)^2)^2 \, (2 \, x + 2)$$

Remark: As an example of the distinction between a function and an expression in this context: in plotting one would do: `plot( D(f), -2..2)`, but `plot( D(f)(x), x = -2..2)`. The D notation provides a nifty way of constructing the tangent approximation to a function. For example, to get the slope and the tangent to the above $f(x)$ at $x = 2$:

```
slope := D(f)(2);
y := f(2) + slope*(x - 2);
```

$$slope := 1800$$
$$y := -\, 2600 + 1800 \, x$$

Exercise 1 Verify each of the above differentiation results by hand.

Analogous to plotting several functions with the `plot` command, we can differentiate more than one function at once:

```
diff( {sin(x)^2, sin(x^2)}, x);
                           2
          {2 cos(x ) x, 2 sin(x) cos(x)}
```

Note: while *Maple* did what we asked, it rearranged the order of the output! So beware.

Exercise 2 Assume that $D \sin x = \cos x$ (this is correct for x is measured in radians). Use the chain rule and this important differentiation result to check the last output.

Although *Maple* can be used to obtain the derivatives of all the functions you know (and many that you haven't met yet), the answer sometimes comes out in a different form from the one you get by hand. Many times this doesn't matter—if *Maple* does the hard work, you should be willing to do a little post simplification yourself (in particular, don't spend 15 minutes forcing *Maple* to do what you can do in 30 seconds!). A "desperation" strategy for checking that two different looking results are actually the same is to **plot** or to table the two forms over some interval. However, it is often worth spending a *small* amount of time trying to use *Maple*'s built-in expression transformers (**simplify**, **factor**, **expand**, and **normal**) to obtain the desired reduction.

IN THE LAB

Exercise 3 Use *Maple* to compute the tangent line to the function

$$f(x) = x^4 - 4x^3 + 3x^2 - 2x + 1$$

at $x = 2$. Check the result by hand.

Exercise 4 Use *Maple* to compute the derivative of $(2x^2 - 1)(x^3 + 2)$. Separately use **expand** and **factor** on the result to get answers more likely to appear "in the back of the book." Use the product rule to check that all your outputs make sense.

Exercise 5 Use *Maple* to compute the derivative of

$$\frac{x}{x-2}$$

and then use **normal** to combine the fractions—**normal** often helps with "quotient rule" derivatives.

Exercise 6 Compute the derivative of $(x + x^2)^5(1 + x^3)^2$ and show that the answer can be written both as

$$5\,x^4 + 30\,x^5 + 70\,x^6 + 96\,x^7 + 135\,x^8 + 210\,x^9 + 231\,x^{10} +$$
$$180\,x^{11} + 156\,x^{12} + 140\,x^{13} + 75\,x^{14} + 16\,x^{15}$$

and as $x^4\,(1+x)^6\,(1 - x + x^2)\,(5 + 5\,x - 5\,x^2 + 16\,x^3)$.

Exercise 7 Use *Maple* to compute the derivative of

$$\frac{1}{1 - 2/x}$$

and then use `simplify` to obtain a cleaner result.

Exercise 8 Pick three nasty looking differentiation problems for which you have answers in the back of your text. Do plain differentiation with *Maple* and then try to find a *Maple* command or commands that will convert the result into a form close enough to the text answer so that the answers can easily be compared. Be sure to state both answers if you don't completely succeed in making them agree. (Be aware that the "back of the book" answer is occasionally wrong!)

AFTER THE LAB

Exercise 9 The results in Exercises 5 and 7 should be identical—why?

BEFORE THE LAB

Prerequisites: Read the material in your text about Newton's method for solving the equation, $f(x) = 0$, paying special attention to the derivation of the Newton iteration,

$$x_{n+1} = x_n - \frac{f(x_n)}{f'(x_n)},$$

for replacing the "current" approximation x_n by the (hopefully) improved approximation x_{n+1}.

Exercise 1 For the specific equation, $x^2 = c$, where c is a given constant, use Newton's method to derive the Babylonian square root iteration scheme of averaging the current guess (say a) and c/a to obtain

$$\frac{1}{2}\left(a + \frac{c}{a}\right)$$

as the new approximation. Using the starting approximation $x = 2$, carry out two iteration steps by hand for the case $c = 5$. Explain your calculations clearly and check the accuracy of your result.

Remark: If $a > \sqrt{c}$, then $c/a < \sqrt{c}$ and *vice versa*. This is the motivation behind the Babylonian method.

We now discuss the implemention of Newton's method in *Maple* in the context of determining a numerical approximation for $\sqrt{5}$. Thus, consider the function definition:

```
f := x -> x^2 - 5;
```

A `plot` (see Figure 5) reveals that $\sqrt{5} \approx 2.2$:

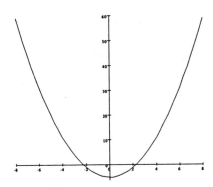

FIG. 5. Result of `plot(f, -8..8);`.

Let's illustrate geometrically how Newton's method generates the first iterate. To get an easily intepretable figure, we use the (poor) starting value of $x = 6$. Figure 6 shows that the tangent at $(6, f(6))$ cuts the x-axis between 6 and the true root, thus producing an improved x-estimate of $\sqrt{5}$ that is about $x = 3.5$.

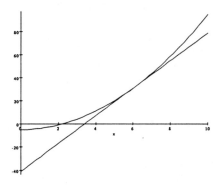

FIG. 6. Geometric demonstration of the first Newton iteration, the poor starting value
$x = 6$ is improved to $x = 3.5$ by this tangent line approximation.

$f'(x)$ **in** *Maple* **is** D(f)(x)

In implementing Newton's method in *Maple* we use the D notation for derivative. Recall
that it works like this:

```
f := x -> x^3:
D(f)(x);
        2
     3 x
```

IN THE LAB

To assess the accuracy of the $\sqrt{5}$ approximations you'll develop shortly, you may find it
convenient to have a highly accurate value available. Here's how to get one to 20 significant
figures:

```
evalf(sqrt(5), 20);
                   2.2360679774997896964
```

We now develop a first *Maple* implementation of Newton's method. First, set the
starting value at $xn = 6.0$:

```
xn := 6.0;   # initialization step
```

Note: It is *important* to use a floating point number such as 6.0 (rather than an integer or
rational such as 6 or 12/2) for the starting value. Otherwise, *Maple* will get bogged down
in trying to do exact calculations.

In place of the mathematical notation that distinguishes the various Newton iterates by
a subscript (i.e., x_n), in our computer implementation, we use the *same* symbol xn for *all*
the iterates—the starting value, the first Newton iterate, and so on. Thus, we compute the
next iterate (the xn on the *left* side of the = sign) from the previous approximation (the
xn's used on the right side) with the statement:

```
xn := xn - f(xn)/D(f)(xn);    # iteration step
```

Exercise 2 Execute the above two commands. You should obtain a value consistent with Figure 6. To get the *next* iterate, just execute the last command again. And so on. Try it for a total of 4 or 5 iterations with the starting value 6.0. Report on the accuracy attained.

Exercise 3 Now try an experiment with taking an even poorer starting value, $x = 15.0$. How many iterations does it takes to converge to $x = 2.23607$?

Exercise 4 One problem with the above method is that we erase each approximation in the process of producing the next one. A simple way to avoid doing this is to use a `for` loop:

```
printlevel := 0:    # so we don't get unwanted printing in the for-loop
xn := 15.0:
for k to 8 do
   xn := xn - f(xn)/ D(f)(xn):
   print( evalf(xn, 16) )
od;
```

To show the approach to convergence, this loop **prints** each of the eight successive iterates to 16 "significant" figures. Try out the loop and then alter it to answer the question: how many iterations does it take to get 10 correct significant figures with the (absolutely insane) starting value 200.0?

Exercise 5 Use an appropriate variant of the above loop to compute the square roots of the seven equally spaced numbers, 0.5, 1.0, 1.5, 2.0, 2.5, 3.5, and 4.0 to 6 significant figures. Use the starting value 1.0 all seven times. (It isn't required, but you may wish to figure out a *double* loop to do it all in one execution.)

Exercise 6 This Lab concludes with the study of the more challenging function

```
f := x -> (x^3 - 2.1*x^2 + x - 2) / (x^6 + 1);
```

a) *Prove* that f has a zero in the interval $[-10, 10]$

b) Graphically show that f has only one zero in this interval. **Hint**: after doing an ordinary `plot` on $[-10, 10]$, repeat using `plot(f(x), x = -10..10, y = -0.1..0.1);` to closely examine the region near the x-axis.

c) Pick a reasonable starting x_n-value, e.g. the closest integer, and do enough Newton iterations to get at least six significant figure precision. Hand in computer output showing all the iterates (`xn` values).

d) Now make `xn` bigger; e.g. `xn = 3.0` or larger. Do several iterations and explain the results.

e) Try `xn` around 0.5 and tell what happens, and why it happens.

f) Try `xn` around -1.0 and tell what happens, and why.

AFTER THE LAB

Exercise 7 In light of your experience with Newton's Method, what caution(s) need to be observed?

Exercise 8 Suppose we want to write a general and efficient routine to find the square root of a given positive number a. As you know, computers deal naturally with binary expansions. One approach is to use this fact to efficiently compute n and b such that $a = 4^n b$ with $0 < b < 4$. Then $\sqrt{a} = 2^n \sqrt{b}$, so we only need a square root routine for numbers between 0 and 4. Since we don't intend to be around when people use our routine, we need to to use a *fixed* starting value x_0 and a *fixed* number of iterations *niter*. Suppose we select the fixed value $x_0 = 1.0$ as we did in Exercise 5. Pick and justify, on the basis of your Lab experience, a likely value for *niter*. How would you make sure that your *niter* was the best possible choice?

Caveat: It is *not* being asserted that the algorithm outlined above is actually the way square roots are computed on any particular hardware. Defining such basic algorithms is a dicey business that depends on the given hardware and on an analysis of how sophisticated a starting guess to use, whether to store precomputed tables, and so forth. So, you see, computing square roots isn't so trivial after all—if *you* are the person designing the algorithm that determines what happens when the calculator button is pressed. If this intrigues you, take a numerical analysis course sometime.

BEFORE THE LAB

Applications of Newton's Method

Law of Reflection.—The law of reflection states that the angle of incidence θ_1 is equal to the angle of reflection θ_2, that is $\theta_1 = \theta_2$ or, equivalently, $\alpha = \beta$ (see Figure 7). Furthermore, if the speed in the medium is constant, then the reflection path from a point P to a point Q consists of two straight line segments meeting at the reflection point as in the Figure.

Exercise 1 Given two points P and Q, we want to determine the location of the point of reflection on the interface (say x measured horizontally from one of the points). As in Figure 7, assume that the two points are separated from each other by a total horizontal distance L and that their vertical distances from the reflecting interface are respectively a and b. Use the law of reflection to derive an expression for x in terms of a, b and L. **Hint:** Find a *simple* equation by using similar triangles. **Check:** For $a = 50$, $b = 25$, and $L = 150$, the correct answer for x is 100.

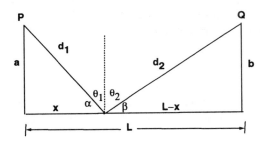

FIG. 7. Geometry of reflection.

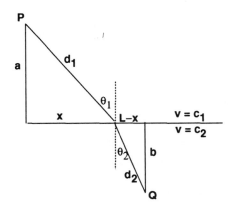

FIG. 8. Geometry of refraction.

Snell's Law.—The law of refraction (often called Snell's law) states that for an interface separating two media, the angle of incidence θ_1 is related to the angle of refraction θ_2, by the relation

$$\frac{\sin \theta_1}{c_1} = \frac{\sin \theta_2}{c_1}$$

(see Figure 8). Here c_1 and c_2 are the respective speeds above and below the interface and θ_1 and θ_2 are again the angles measured from the normal. Once more if the speeds are assumed constant, then the refraction path consists of two straight line segments meeting at the refraction point.

In contrast to the reflection case where the reflection point can be found explicitly, the location of the refracting point x involves a fourth degree polynomial:

$$(c_2^2 - c_1^2)x^4 - 2L(c_2^2 - c_1^2)x^3 + (L^2(c_2^2 - c_1^2) + c_2^2 b^2 - c_1^2 a^2)x^2 + 2Lc_1^2 a^2 x - L^2 c_1^2 a^2 = 0 \quad (2)$$

In real applications, this "ray tracing" problem becomes even more difficult (and often takes super-computing capability) because many layers are involved. Moreover, in many applications, it is necessary to deal with complications such as allowing the speed to vary in each layer or considering distinct wave types simultaneously (e.g., elastic "P-waves" and "S-waves"), etc. Ray tracing plays an important part in many fields ranging from geophysics to creating cinematic special effects.

Asteroid Problem.—The previous work on this problem showed that the radius r of the asteroid is governed by the equation

$$\cos \frac{s}{r} = \frac{r}{r + h},$$

where s is the distance the traveling astronaut walks and h is the length of the rod that she holds vertically for the stationary ground-level astronaut to sight on.

Exercise 2 Recall that for the particular values $s = 400 \, \text{m}$ and $h = 1 \, \text{m}$, you've shown that $r \approx 80000 \, \text{m}$. To avoid "wasting" the 5 significant figures involved in writing "80000," make the substitution $r = R + x$ in the asteroid equation (here $R = 80000$, but for goodness sakes don't write out "80000" three times, write "R").

IN THE LAB

Snell's Law continued.—

Exercise 3 Show that for the values $a = b = 1$, $L = 4$, $c_1 = 1$, and $c_2 = 1/2$, Equation 2 for the location of the refraction point reduces to

$$3x^4 - 24x^3 + 51x^2 - 32x + 64 = 0.$$

This latter equation actually has *two* real roots:

a) Hand in a graph that shows their approximate locations.

b) Explain why only one of the roots is relevant to the refraction problem.

c) Find the relevant zero to six significant figures. Show the result of each Newton iteration.

d) Provide a graph showing an interval including the relevant zero and the tangent line from your starting $(x, f(x))$ intersecting the x-axis at the first iterate.

Exercise 4 Using Snell's Law, Figure 8 implies

$$\frac{1}{c_1}\frac{x}{d_1} = \frac{1}{c_2}\frac{L-x}{d_2}.$$

Cross multiplying, squaring and transposing gives

$$(c_2\, x\, d_2)^2 = (c_1(L-x)d_1)^2.$$

Use *Maple* to simplify this expression and hence derive Equation 2.

Asteroid Problem continued.—

Exercise 5 So far, we've applied Newton's method to polynomials, but it applies just as well to transcendental equations such as the one you obtained in Exercise 2. Obtain the relevant root x (not r!) to 4 significant figures with Newton's method. State the corresponding solution for the physical quantity r.

Exercise 6 When you previously found the "physically sensible" solution to the asteroid problem you may have noticed that there were a lot (infinitely many, actually) of roots r in the interval $[0, 400]$. Find the largest two roots (r-values) that are less than $s = 400$ to 6 significant figures and explain what would have happened to the astronauts if the asteroid really had a size corresponding to the radii given by these two smaller roots. Similarly, explain what happens for an asteroid with radius given by the yet smaller roots.

AFTER THE LAB

Exercise 7 Derive the law of reflection from Fermat's principle of least time: light will follow the path that minimizes the time. You may assume that the paths are straight lines. **Hints:** Since the speed is constant, this is equivalent to minimizing $d_1 + d_2$ in Figure 7. After taking a derivative, interpret the result in terms of the angles α and β. You could also determine x from this equation, but hopefully you found a simpler way in Exercise 1.

BEFORE THE LAB

Nothing to do beforehand this time. We know that Newton's method usually converges quickly, if we start "sufficiently close" to a simple root. In this Project, we examine some problems with the method. Don't get paranoid, we are going out of our way to create difficulties—in most cases, Newton's method is a wonderful tool.

IN THE LAB

Exercise 1 Run the usual Newton iteration loop on the function

$$f(x) = 27x^6 - 216x^5 + 387x^4 - 440x^3 + 408x^2 - 224x + 48$$

with starting value 1.0.

a) How many Newton iterations does it take to get 5 significant figure accuracy?

b) Plot f for $-2 < x < 2$ and try to explain why the Newton method is unusually slow.

Chaos Theory.—In the remaining problems, we apply Newton's method to the simple function $f(x) = x^3 - x = x(x - 1)(x + 1)$ with the roots $x = -1$, 0 and 1. If the initial value x_0 is chosen close to any one of these three roots, the Newton iterations converge quickly to that root. However, if one selects x_0 on a certain interval (it isn't near any root), strange things happen. This behavior is related to an area of mathematics called "chaos" theory. Chaos theory has received great attention during the past 10 years and there was even a best seller about it[1]. The term "chaos" arises from the fact that for some problems (e.g., weather prediction), *small* changes in the starting conditions make *large* changes in the long term behavior. Thus, long term prediction becomes impossible, despite knowing the exact equations.

Exercise 2 Two important x-values in our study are $s3 = 1/\sqrt{3}$ and $s5 = 1/\sqrt{5}$. One would *not* want to pick x_0 as either $\pm s3$. Why not? Also, what happens if one picks $x_0 = s5$? That is, what are $x_1, x_2 \ldots$?

It is useful to have **plots** of f in both the critical region $[-1, 1]$ and on a larger scale, say $[-4, 4]$.

Exercise 3

a) Explain from the plots why you would *expect* that an initial value $x_0 > s3$ would lead to $x_n \to 1.0$ as $n \to \infty$. (And by symmetry, $x_0 < -s3$ leads to $x_n \to -1.0$ as $n \to \infty$).

b) Similarly, explain why if $|x_0| < s5$, you would expect $x_n \to 0$.

c) Numerically verify the assertions in parts 1 and 2.

---

[1] *Chaos* by James Gleick, Viking-Penguin Inc., 1987.

Exercise 4 (The "chaos" part). Provided here is a list of starting values x_0 lying between $s5$ and $s3$ and getting closer and closer to $s5$. Using each starting value, run 15 or so Newton iterations and carefully describe what you observe (including any sign changes in the Newton iterates).

   a) use $x_0 = 0.448955$

   b) use $x_0 = 0.447503$

   c) use $x_0 = 0.447262$

   d) use $x_0 = 0.447222$

   e) use $x_0 = 0.447215$

   f) use $x_0 = 0.447213$ (whoops!)

AFTER THE LAB

Exercise 5 Discuss the results obtained in Exercise 4 with respect to "sensitivity" to the choice of initial value.

Exercise 6 Suppose that it is known that for a certain function f, a unique root exists on the interval $[a, b]$. We've seen that in some unusual cases, Newton's method can either fail or can converge too slowly. On the other hand, settling for a "safe" method like bisection *all the time* is too expensive. How could we automatically monitor when Newton's method is in "trouble"? Casually discuss "rescuing" Newton's method in such cases.

Finding Roots with Maple (optional)

BEFORE THE LAB

Now that we understand how Newton's Method works, we look at several equation solvers that are built in to *Maple*—of course, Newton's method is a key element in their construction!

We begin by distinguishing the limitations and strengths of the equation solving commands: `solve`, and `fsolve`. `solve` can handle polynomial equations (of degree less than 5) and a few other types of simple equations (e.g. that can be solved using elementary functions and their inverses).

`solve` attempts to give *exact* solutions, while `fsolve` gives only *approximate*, numerical solutions. `solve` cannot exactly solve all polynomial equations (there is no mathematical method to do this for equations of degree 5 and higher). However, `solve` can sometimes handle equations with *parameters*, whereas `fsolve` cannot. `fsolve` often requires additional information—typically an interval in which you are looking for the root. This is especially true if the equation is complicated or has more than one root.

The quadratic equation

If we want to obtain the familiar solution formula for the quadratic equation,

$$ax^2 + bx + c = 0,$$

then our only chance is `solve` since only it can handle parameters like a, b and c;

```
quadeqn := a*x^2 + b*x + c = 0:
solutions := solve(quadeqn, x);
                                 2    1/2              2           1/2
                      - b + (b  - 4 a c)       - b - (b  - 4 a c)
       solutions := 1/2 --------------------, 1/2 --------------------
                               a                          a
```

There are a couple of things to observe in this simple example. Notice we saved the results in a variable called 'solutions'; hence we can access the *two* roots to the equation. For example, we can see if the roots really do solve the quadratic equation. Let's check the first root, called 'solutions[1],' by substituting it back into the equation:

```
subs(x = solutions[1], quadeqn);
             2           1/2 2                    2           1/2
       (- b + (b  - 4 a c)   )          b (- b + (b  - 4 a c)   )
   1/4 ------------------------- + 1/2 ------------------------- + c = 0
                 a                                a
```

```
simplify(");
                              0 = 0
```

The output tells us that the first root actually worked.

A cubic equation

Now we turn to the numerical solution of the cubic equation $x^3 - 3x^2 + 1 = 0$. While `solve` would certainly work here, perhaps the most efficient way is to use `fsolve`:

```
Digits := 10:
lhs := x^3 - 3*x^2 + 1:
solutions := fsolve(lhs = 0, x);
```

This time, we used the variable `lhs` to name the left hand side of the equation and put the '=0' inside the `fsolve` command. To check the accuracy of the *second* solution:

```
subs(x = solutions[2], lhs);
```
$$.3*10^{-9}$$

The error is "in the noise"; that is, within the 10 digit accuracy asked for.

Alternately, since this is only a cubic, you could use the less efficient approach of asking for the exact solutions via `solve`. Since the roots are not particularly 'nice,' one will get nasty looking results in expressing the exact answers. However, following with a `evalf(")` will convert to numerical form. Try it.

`solve` won't work on most non-polynomial equations, for example, it fails on the "asteroid" equation. Try it. But `fsolve` works fine, once we have the desired root bracketed. Notice in the `fsolve` command how the search interval is indicated.

```
s := 400: h := 1:
equation := cos(s/r) = r/(r + h):
solution := fsolve(equation, r, 80000..80001);
                    solution := 80000.83332
subs(r = solution, equation);
            cos(.004999947920) = .9999875003
```

```
evalf(");
            .9999875003 = .9999875003
```

Here we first got the answer, then checked the accuracy.

Exercise 1 A 100-ft tree stands 20 feet from a 10-ft fence. The tree is then "broken" at a height of x feet, so that it just grazes the top of the fence and touches the ground on the other side of the fence with its tip (see Figure 9). Derive the equation $x^3 - 68x^2 + 1100x - 5000 = 0$ for x.

IN THE LAB

Exercise 2 Use `solve` to get exact answers and then use `fsolve` to get numerical values for the quadratic equation, $x^2 - 3x + 4 = 0$.

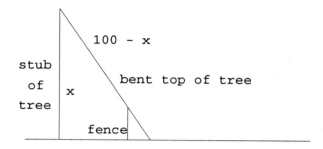

FIG. 9. Diagram for Exercise 1.

Exercise 3 Does `solve` work on the polynomial equation $x^5 - 3x + 4 = 0$?
Note: in *Maple* when solving polynomial equations one sometimes gets a response
`RootOf('‘something’')`.
If so, follow up with the command `allvalues(")` to get numerical results.
In any case, find all real roots.

Exercise 4 Solve the "broken tree" equation using `fsolve` and decide which roots make physical sense.

Exercise 5 Use *Maple* to solve the cubic equation, $x^3 - 2ax^2 - 4x + 8a = 0$, and to check the answers. **Remark**: This equation is special—most problems like this have answers that are "17 pages long" and hence the exact solution is of little practical use.

Exercise 6 Plot $\tan x + x$ on $[0, 4\pi]$ to get crude values for its positive zeroes near the origin (note: you may want to restrict the *range* of the output in plotting). Use `fsolve` to get accurate values for the first 4 positive roots of the equation $\tan x + x = 0$ and check the accuracy of your solutions.

AFTER THE LAB

Nothing this time.

Area

Introduction

A fundamental approach to finding the area of an object with general shape is to approximate it by sums of rectangles (since we *know* the area of a rectangle). For the area between the x-axis and a positive function $f(x)$ on the interval $[a, b]$, your text introduces the idea of a systematic Riemann sum approximation, which may be written as

$$\text{Area} \approx R = \sum_{i=0}^{n-1} f(x_i^*) \Delta x_i, \qquad (3)$$

where x_i^* is a point in the interval $[x_i, x_{i+1}]$ and Δx_i is the length of this interval (i.e., $x_{i+1} - x_i$). In this project we will use this idea to approximate some interesting areas. We choose the specific Riemann sum obtained using *equal* subdivisions $\Delta x_i = \Delta x = (b - a)/n$ and the *midpoints* $x_i^* = a + (i + 1/2)\Delta x$, so that the sum approximating the area is given by

$$\text{Area} \approx R = \sum_{i=0}^{n-1} f\left(a + (i + 1/2)\Delta x\right) \Delta x, \qquad \Delta x = (b - a)/n.$$

Figure 10 shows an approximation of this type for a small value of n.

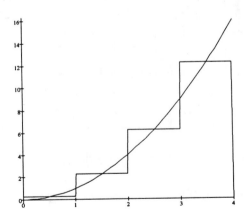

FIG. 10. Function and rectangular area approximation.

Exercise 1 Equation 3 is often written as

$$R = \sum_{i=1}^{n} f(x_i^*) \Delta x_i,$$

where now x_i^* is a point in the interval $[x_{i-1}, x_i]$ and Δx_i is the length of this interval (i.e., $x_i - x_{i-1}$). Explain why this equation expresses the same quantity.

Area of a circle

You know that the area of a circle of radius r is $A = \pi r^2$, where $\pi \approx 3.14$. But until now, you have probably just accepted this dictum. It is easy to argue from similarity that the area is *proportional* to r^2. And it is easy to give the constant of proportionality a *name* like π. What is *not* easy is to give a method for accurately computing π. But before pursuing that, let's start with the crude approach of simply drawing a circle on a grid and counting the squares enclosed (sometimes we have to get results in the absence of a good theoretical framework).

Exercise 2 Count the squares totally enclosed by the circle in Figure 11 and add in an estimate of the portion of squares on the boundary that are also inside to get an estimate of the area of the circle. Use your results to estimate π.

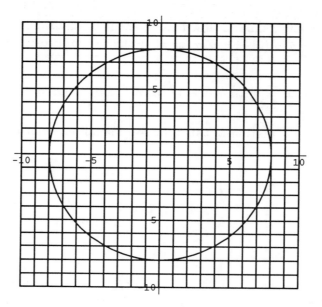

FIG. 11. Circle with superimposed grid.

Exercise 3 Use Figure 12 to obtain an estimate of the area of the ellipse shown there.

Exercise 4 We have implicitly *defined* the constant π as the area of the unit circle. Use *only* this definition to show that π is exactly four times the area under the curve $y = \sqrt{1 - x^2}$ on the interval $[0, 1]$.

The *Maple* sum Command

Although sums can be evaluated by coding a **for** loop, *Maple* provides a "higher level" construct for this common task. For example, $2^2 + 3^2 + 4^2 + 5^2 + 6^2$ can be computed as

```
sum(i^2, i = 2..6);
```
$$90$$

Exercise 5 Make sure you understand what **sum** does by checking this result by hand.

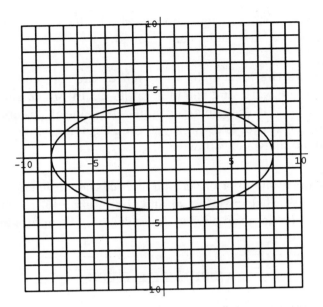

FIG. 12. Ellipse with superimposed grid.

IN THE LAB

Exercise 6 Use code like:

```
n := 4:
sum(i^3, i = 1..n);
```

to evaluate sums of cubes for the cases $n = 4, 16, 64, 256$, and 1024 (i.e., $n = 4^k$, for $k = 1$, 2, 3, 4, and 5).

Exercise 7 Make a guess about what `product(i^2, i = 2..6)` should produce and then check your guess by running the command. Finally, write a clear explanation of how the `product` command works.

The Riemann Sum in *Maple*

We now use the `sum` command to compute the Riemann sum based on the equal subdivisions and midpoints defined above in the Introduction.

For our circle problem, set up the function and limits like this:

```
f := x -> sqrt(1 - x^2):
a := 0:  b := 1.0:
```

Next we write the `sum` command to form the Riemann sum. Note: Since $\Delta x = $ `delx` is a constant, we take it outside the `sum`.

```
n := 5:  delx := (b - a)/n:
Rsum := delx * sum( f(a + (i+.5)*delx), i = 0..n-1):
approxPi := 4 * Rsum;
                    approxPi := 3.171987824
```

Exercise 8 After repeating the previous run as a check, determine n to verify the approximation $\pi \approx 3.14$ (correct to three significant figures).

Exercise 9 Approximate the area under one arch of the sine curve to three significant figures. That is, use $f(x) = \sin x$ and the interval $[0, \pi]$. In *Maple* $\sin x$ is written `sin(x)` and π is written `Pi`. **Note**: With calculus, the numerical value of this area can be determined *exactly*. In fact, the exact answer is 2. So if you aren't getting something around 2, take a deep breath and think about what you are doing.

Exercise 10 Approximate the area under one arch of the curve $\sin(\sin x)$ to three significant figures. That is, use $f(x) = \sin(\sin x)$ and the interval $[0, \pi]$. **Note**: the numerical value of this area can*not* be determined *exactly*. It can only be approximated with methods such as the one you are using.

AFTER THE LAB

Exercise 11 Search your text for a formula that will allow you to check the results you obtained in Exercise 6.

Exercise 12 The idea of approximating areas with curved boundaries by sums of simple areas is attributed to Archimedes. Find out when Archimedes lived and give at least three interesting accomplishments of his.

Introduction

If the function f has an explicit antiderivative F in terms of elementary functions, then the Fundamental Theorem of Calculus tells us that

$$\int_a^b f(x)\,dx = F(b) - F(a),$$

so that evaluating the definite integral of f can be achieved by merely evaluating F at a and b. For a continuous f, the Existence Theorem and Fundamental Theorem together guarantee that the definite integral and antiderivative exist (for example, $F(x) = \int_a^x f(t)\,dt$). However, for many elementary functions f, the antiderivative F is not itself an elementary function that we can evaluate directly and hence the Fundamental Theorem does not provide a means for evaluation of the definite integral. In such cases, we can still get accurate numerical approximations to the definite integral—the Existence Theorem guarantees this! The numerical approximation of integrals is a major topic in its own right and we can only scratch the surface in a first course in calculus. In this project, we examine the first step—direct use of the Riemann sums that are used in the proof of the Existence Theorem.

In this and the next Project, we examine the *quality* of some basic numerical approximations, that is, we find out how good a job they do in estimating the exact integral. In this Project, you will see that some Riemann sums provide better approximations than others, even when using the same number of sub-intervals.

Exercise 1 Use the Fundamental Theorem to check that

$$F(x) = \int_a^x f(t)\,dt$$

is an antiderivative of $f(x)$.

Exercise 2 Show that the exact area between the x-axis and the function $f(x) = x^4 + 1$ on the interval $[0, 5]$ is 630.

Exercise 3 Explain the geometrical meaning of the curve $y = \sqrt{100 - x^2}$ on the interval $[0, 10]$ and thus evaluate exactly the definite integral,

$$I = \int_0^{10} \sqrt{100 - x^2}\,dx.$$

IN THE LAB

The Riemann Sum in *Maple*

As a "warm-up," we compute some Riemann sums for $f(x) = x^4 + 1$ on $[0, 5]$. We start with $n = 4$ and will increase n and observe what happens.

Set up the function and limits like this:

```
f := x -> x^4 + 1:
a := 0: b := 5.0:
```

Next, we use the **sum** command to form the particular Riemann sum in which the *left* endpoint is used on each sub-interval. Study this command carefully. In particular, note that when $n = 5$, i takes on the values 0, 1, 2, 3 and 4. Hence f is evaluated at $x = 0$, Δx, $2\Delta x$, $3\Delta x$ and $4\Delta x$. Since $\Delta x = $ **delx** is a constant, we take it outside the **sum**:

```
n := 5:  delx := (b - a)/n:
Rsum :=  delx * sum(f(a + i*delx), i = 0..n-1);
                        Rsum := 359.0000000
```

Notice that the answer is awful!

Exercise 4 After repeating the above run as a check, compute the Riemann sums for $n = 10$ and $n = 100$.

We now improve the above Riemann sum code, which is so far restricted to using the left endpoint of each sub-interval. Introduce a parameter p (standing for "proportion"), where p is a value between 0 and 1. Notice that when $p = 0$ the new **sum** command below acts just like the above **sum** command. But when $p = 1$, we get the Riemann sum using the *right* endpoint of each sub-interval. The intermediate choice $p = 0.5$ gives the "midpoint rule" (as used in the Area project). So consider the following code in which we first set $p = 1$:

```
exact := 630:
n := 10:  delx := (b - a)/n:
p := 1.0:
Rsum :=  delx * sum(f(a + (i+p)*delx), i = 0..n-1);
```

Exercise 5 Repeat Exercise 4, this time using *right*-endpoint sums. Contrast the results.

Exercise 6 Add the line **exact - Rsum** to the improved Riemann sum code and make your study systematic by filling in an *error* table with column headings like this:

```
  left-endpoint (p = 0)     midpoint (p = 0.5)      right-endpoint (p = 1)
```

and rows labeled by $n =10$, 100, and 1000. The nine entries in the table should be the errors in the corresponding Riemann sum. (It is not necessary to do anything fancy unless you want to, just get the nine values and write them in a table by hand.)

Exercise 7 Do the analog of the previous problem for the function and interval in Exercise 3.

AFTER THE LAB

Exercise 8 The errors in numerical approximations tend to be proportional to $1/n$, or to $1/n^2$, or to $1/n^3$ and so on, where n is the number of sub-intervals. Try to detect this pattern: on the basis of the tables you made in Exercises 6 and 7, find the power of n for each of the values of p. **Hint**: When n increases by a factor of 10 (say from 10 to 100), then if the error decreases by a factor of 10 it is behaving like $1/n$. On the other hand, if the error decreases by a factor of 100, then it is behaving like $1/n^2$, etc. Fractional powers are possible.

Exercise 9 Can you give an intuitive explanation of why Riemann sums with $p = 1/2$ should usually have smaller errors than those with $p = 0$ or 1?

Exercise 10 Let

$$f(x) = e^{\sin \sqrt{x^{12} + \cos^2 x}}.$$

Does $\int_0^1 f(x)\,dx$ exist? Does the Fundamental Theorem help you to evaluate it? If so, how? Could you get a good numerical approximation? If so, how?

BEFORE THE LAB

This project reviews the connection between the area function and the Fundamental Theorem and then implements three basic numerical integration methods in *Maple*: the trapezoid rule, the midpoint rule and Simpson's rule.

The Area Function

We study the area function $A(x)$ associated with a continuous function $f(x)$. $A(x)$ is defined as the integral of f from a given, fixed left endpoint a to the variable point x:

$$A(x) = \int_a^x f(t)\,dt.$$

Exercise 1 Using geometry (i.e, no calculus allowed!), construct the area function for the following functions:

a) $f(x) = c,$ $c = \text{constant}$

b) $f(x) = cx,$ $c = \text{constant}$

Using the `int` Command to Construct the Area Function

In *Maple*, the area function is constructed like this:

```
A := x -> int( f(t), t = a..x );
```

Remark: We are living a bit dangerously, since a and f are left as implicit variables—that is, they have whatever values you last set. Initially, they are not set at all, so if you ask for `A(x)`, you just get *Maple*'s version of the symbolic form echoed back:

```
A(x);
```

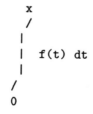

```
      x
     /
    |
    |   f(t) dt
    |
   /
   0
```

Exercise 2 Explain this input/output:

```
diff( A(x), x );
```

$$f(x)$$

Later, when we set f and a, we get an updated response:

```
f := x -> 5;
a := 1:
```

$$f := 5$$

```
A(x);
```

$$5x - 5$$

Exercise 3 Show that the previous output makes mathematical sense.

If you ever need a clean slate for, e.g., f and a:

```
f := 'f':   a := 'a':
```

Now, if you define `A(x)` as above you would just get the integral expression as before.

The Basic Numerical Integration Methods in *Maple*

In a previous project, we gave code for computing Riemann sums. Here we repeat that code, but now in the form of a **procedure**, *Maple*'s version of a subroutine:

```
Rsum := proc(f, a, b, n, p)
   delx := (b-a)/n;
   delx * sum ( f(a + (i+p)*delx), i = 0..n-1 )
end;
```

The parameter p controls the kind of Riemann sum formed: $p = 0$ gives the Left Rectangle Rule, $p = 1$ gives the Right Rectangle Rule and $p = 0.5$ gives the Midpoint Rule. Valid, but less popular, Riemann sums are given for other choices of p between 0 and 1. Here's how to use `Rsum`:

```
f := x -> sin(x):
a := 0.0: b := evalf(Pi):
n := 5:  p := .5:
Rsum(f, a, b, n, p);
```
$$2.033281477$$

In this project, we compare three basic numerical integration methods (by the way, there are entire texts devoted solely to developing and analyzing numerical integration methods!). The first is the *Midpoint rule*. This is simply `RSum` with $p = 0.5$, so we obtain the procedure:

```
Mid := proc(f, a, b, n)
   Rsum(f, a, b, n, 0.5)
end;
```

The Trapezoid rule can be defined as the average of the `Rsums` with $p = 0$ and $p = 1$ (Left and Right Rectangle rules), so

```
Trap := proc(f, a, b, n)
   0.5 * ( Rsum(f, a, b, n, 0) + Rsum(f, a, b, n, 1) )
end;
```

Finally, Simpson's rule can be defined as the weighted average of the Midpoint rule and the Trapezoid rule with the former having twice the weight of the latter, so that

```
Simp := proc(f, a, b, n)
   ( Trap(f, a, b, n, 0) + 2*Mid(f, a, b, n, 1) ) / 3.0
end;
```

Here's a typical usage for these modules:

```
f := x -> sin(x):
a := 0.0: b := evalf(Pi):
n := 5:
Simp(f, a, b, n);
```

$$2.000109518$$

For easy reference, we give the explicit formulas with error term for the three rules examined in this project. With the usual $\Delta x = (b-a)/n$, we use the notation, $f_i \equiv f(a + i\Delta x)$, to write:

$$
\begin{array}{lll}
\text{Mid} & : & \Delta x(f_{\frac{1}{2}} + f_{\frac{3}{2}} + \cdots + f_{n-\frac{1}{2}}) - \dfrac{f''(\theta_M)(b-a)^3}{24n^2} \\[2ex]
\text{Trap} & : & \Delta x(\dfrac{f_0}{2} + f_1 + f_2 + \cdots + f_{n-1} + \dfrac{f_n}{2}) + \dfrac{f''(\theta_T)(b-a)^3}{12n^2} \qquad (4) \\[2ex]
\text{Simp} & : & \dfrac{\Delta x}{3}(f_0 + 4f_1 + 2f_2 + 4f_3 + \cdots + 2f_{n-2} + 4f_{n-1} + f_n) + \dfrac{f^{(4)}(\theta_S)(b-a)^5}{180n^4}.
\end{array}
$$

In each case, the final derivative term represents the error in that rule and the θ's denote unknown points in the interval $[a, b]$, just as in the mean value theorem.

Exercise 4 It comes as a bit of a surprise that Simpson's Rule gives the *exact* result for cubic polynomials. Use the theory expressed in Equation 4 to explain this.

IN THE LAB

Exercise 5 Run the following commands:

```
A := x -> int( f(t), t = a..x );
f := x -> sin(x):
a := 0.0: b := evalf(Pi/2):
u := f(x):  y := A(x):  z := D(A)(x):
```

and then **plot** u, y and z on the same graph. Explain the result.

Exercise 6 We study the errors obtained by the Midpoint, Trapezoid and Simpson rules as a function of the number of sub-intervals, n. To make the study easier, we provide you a loop that computes these *errors*. These errors are called **MidE**, etc. Note that n increases as powers of 2 (as $k = 1, 2, ..., 7$ in the **for** loop). This will make it easy for you to see the "order of convergence" of the various rules. So execute the following:

```
f := x -> sin(x):
a := 0.0: b := evalf(Pi):
exact := 2.0:

printlevel := 0:
lprint( 'x          MidE          TrapE          SimpE');
for k to 7 do
   n := 2^k:
   lprint( n,
           exact - Mid(f, a, b, n),
           exact - Trap(f, a, b, n),
           exact - Simp(f, a, b, n) )
od;
```

Remark: You have to execute the integration modules before the above loop will work.

Exercise 7 Repeat the last exercise with $f(x) = \sin(\sin(x))$. There is no useful antiderivative for this function, so you cannot get an exact answer and numerical approximations are required. *Maple*'s `int` command followed by `evalf(")` computes most definite integrals accurately. So, to get a good approximation to the exact value execute:

```
f := x -> sin(sin(x)):
a := 0:  b := evalf(Pi):
int( f(x), x = a..b );
exact := evalf(");
```

Exercise 8 Do it one more time with $f(x) = \sqrt{(4 - x^2)}$ and with $a = 0$ and $b = 2$. For this function, you can get the exact answer from knowledge of the area of a circle.

AFTER THE LAB

Exercise 9 In regard to the area function $A(x)$ defined above:

a) What is $\dfrac{dA(x)}{dx}$?

b) What is the value of $A(a)$, where a is the left endpoint? Why?

c) Corresponding to a continuous function $f(x)$, suppose that a clever friend has found an antiderivative, call it $F(x)$. How is her $F(x)$ related to $A(x)$?

Exercise 10 The errors in numerical integration generally behave like C/n^k. From your results for Exercise 6, determine k and C for each of the three integration rules examined there. (Suggestion: for various technical reasons, you may want to compute k and C by comparing numerical results for, say, $n = 16$ and 32, rather than larger n). Also, note the relation between the errors in the Trapezoid and Midpoint rules. Answer carefully: do your results agree with Equation 4?

Exercise 11 Repeat the last exercise using the results you obtained in Exercise 7.

Exercise 12 And repeat again using the results you obtained in Exercise 8.

BEFORE THE LAB

Exercise 1 We know that the equation $x^2 + y^2 = r^2$ represents a circle of radius r. Use this fact to justify the formula:

$$A_{circle} = 4r \int_0^r \sqrt{1 - (x/r)^2}\, dx = \pi r^2 \qquad (5)$$

for the area of this circle.

Exercise 2 Similarly, the equation

$$\left(\frac{x}{a}\right)^2 + \left(\frac{y}{b}\right)^2 = 1$$

represents an ellipse with axes a and b. Justify both of the formulas:

$$A_{ellipse} = 4b \int_0^a \sqrt{1 - (x/a)^2}\, dx \qquad (6)$$

$$= 4a \int_0^b \sqrt{1 - (y/b)^2}\, dy \qquad (7)$$

for the area of this ellipse.

Exercise 3 Make the substitution $u = x/a$ in Equation 6 and use Equation 5 to compute $A_{ellipse}$ explicitly. Check your result in each of the following two ways: (1) derive the result again using Equation 7 and (2) show that your result specializes correctly in the case when the ellipse becomes a circle of radius r.

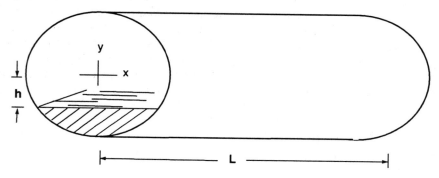

FIG. 13. Partially filled tank.

The Tank Problem

Figure 13 shows an underground fuel tank (e.g., as at a service station) in the shape of an elliptical cylinder. The tank is lying on its side (i.e., the cylinder is horizontal) and has length L ft. The elliptical cross-section has axes a ft and b ft. The owner has only a crude way of measuring the contents: he lowers a stick down into the tank and measures how high the fuel level is on the stick. You are to help him convert this measurement, call it h (in ft), into cubic ft of fuel in the tank.

Exercise 4 Derive the following formula for the volume $V = V(h)$, where $-b \leq h \leq b$:

$$V = 2La \int_{-b}^{h} \sqrt{1 - (y/b)^2} \, dy \tag{8}$$

Hint: the volume is an area times the length of the tank.

Using the `int` Command to Evaluate Definite Integrals

In *Maple*, the definite integral is written as: `int(f(x), x = a..b);`. As an explicit example, the area under x^3 between $x = 5$ and $x = 10$ is written as:
`int(x^3, x = 5..10);`

IN THE LAB

Exercise 5 Use *Maple* to check your evaluation of the integrals in Equations 5, 6 and 7.

Exercise 6 In reference to the tank volume in Equation 8, use the `int` command to obtain the volume as a function of h with general parameters a, b and L.

Exercise 7 Continuing with the tank problem: for the specific parameter values, $L = 20$, $a = 10$ and $b = 5$, make a table with two columns, containing respectively h and $V(h)$ from $h = -5$ to $h = 5$ with unit spacing. Coding hint:

```
V := h -> ...
for k to 5 do
  ...
od;
```

Exercise 8 Your friend would like to put a mark on the stick as a warning to indicate when only 500 ft^3 are left. Using the same numerical parameter values as in Exercise 7, figure out the height h at which the stick should be marked. **Hints**: Use Newton's method to solve the equation $V(h) = 500$. Newton's method tries to find a solution to $f(x) = 0$, what is the f for this problem? Newton's method requires a starting guess, how can you get a good one? Newton's method requires a derivative—in this problem the derivative is easy to get, why? Finally, express your answer as the distance from the *bottom* of the stick. Coding hint:

```
L := 20:  a := 10:  b := 5:
V := h -> ...
Vprime := h -> ...
hn := ...
for h from -5 to 5 do
   hn := ...
od;
```

AFTER THE LAB

Exercise 9 Explain why Exercise 5 implies that the total volume of the cylindrical tank is πabL and use this result to check your output in Exercise 7 for the values $h = -5$, $h = 0$, and $h = 5$.

BEFORE THE LAB

Introduction

It turns out that if y is an ordinary elementary function, only rarely does the arclength integrand $\sqrt{1 + y'^2}$ have an elementary antiderivative. Hence, the textbook exercises on computing arclength integrals tend to be artificial. In this project, we'll investigate what's behind some of the weird functions that appear in textbooks and then use *Maple* to evaluate some arclengths of natural interest. We begin by defining a *Maple* function to compute the differential arclength ds corresponding to a given function y (notice that we leave out the dx that appears in the mathematical expression for differential arclength):

```
ds := (y,x) -> sqrt(1 + D(y)(x)^2);
```

Using the Fundamental Theorem

Certain specific elementary functions y have differential arclengths $ds = \sqrt{1 + y'^2}$ with elementary function antiderivatives. In this case, the Fundamental Theorem of Calculus (FTC) can be used to compute the arclength. For example, finding the arclength under the parabola $y = x^2$ involves integrating the square root of a quadratic:

```
y := x -> x^2:
dsParabola := ds(y,x);
                                        2 1/2
                  dsParabola := (1 + 4 x )
```

It turns out that integrands involving "at worst" square roots of quadratics have elementary function antiderivatives, while integrands involving square roots of cubic or higher polynomials generally do not. We will treat integration of the square root of quadratics later in this semester (look up "trigonometric substitutions," if you are too curious to wait). For now, we appeal to *Maple* to get the antiderivative for this particular integrand:

```
sParabola := int( dsParabola, x );
                          2 1/2                        2 1/2
        sParabola := 1/2 x (1 + 4 x )    + 1/4 ln(2 x + (1 + 4 x )   )
```

So, the arclength from (say) 0 to 1 can be computed exactly by the FTC as

```
subs( x=1, sParabola ) - subs( x=0, sParabola );
                  1/2                1/2
          1/2 5      + 1/4 ln(2 + 5   ) - 1/4 ln(1)
evalf(");
                          1.478942858
```

But for most elementary functions $y(x)$ the FTC can*not* be used because an antiderivative in terms of elementary functions doesn't exist. For example, just replace x^2 by x^3:

```
y := x -> x^3:
int( ds(y,x), x );
                                          /
                    4 1/2                 |        1
          1/3 x (1 + 9 x )     + 2/3      | ------------- dx
                                          |        4 1/2
                                          /   (1 + 9 x )
```

Arclength (optional)

Notice that *Maple*'s "answer" involves an unevaluated integral; this is the way *Maple* tells us that an elementary function antiderivative couldn't be found.

Exercise 1 Assume that integrands involving "at worst" square roots of quadratics have elementary function antiderivatives. That is, assume that integrands involving $\sqrt{a + bx^n}$ with $n = 2, 1, 0, -1$ and -2 have elementary antiderivatives. Determine which functions of the form $y = x^p$ have differential arclengths with elementary antiderivatives. Two of the powers p correspond to lines—which two? Look at the problems in your calculus book and identify the arclength exercises involving the non-trivial powers p you have found.

Exercise 2 Since so few functions we "meet in the street" lead to integrands with elementary function antiderivatives, textbook authors concoct exotic functions that "work out." Typical examples are:

a) $y = \dfrac{x^3}{3} + \dfrac{1}{4x}$

b) $y = \ln|x| - \dfrac{x^2}{8}$

Verify by hand that each of the above expressions for y miraculously leads to an arclength integral free of square roots.

A person with a pinch of curiosity just has to be wondering: "How in the world do they come up with these weird examples?" From the experience gained in solving the last problem, it emerges that the essential element is that $1 + y'^2$ turns out to be a perfect square.

Here's the secret: Compare the identity

$$(f + g)^2 = (f - g)^2 + 4fg$$

with the desired relation

$$s'^2 = y'^2 + 1.$$

This motivates identifying s' with $f + g$ and y' with $f - g$. The identification is complete if we set $4fg = 1$. Hence take $g = 1/(4f)$ to obtain:

$$y' = f - \frac{1}{4f},$$

$$s' = f + \frac{1}{4f},$$

or

$$y = D^{-1}\left(f - \frac{1}{4f}\right), \tag{9}$$

$$s = D^{-1}\left(f + \frac{1}{4f}\right), \tag{10}$$

where D^{-1} denotes antidifferentiation.

Thus, to create "doable" textbook arclength problems, we need merely pick an f for which both f and $1/f$ have elementary antiderivatives. Then Equation 9 provides the textbook problem and Equation 10 provides the corresponding arclength function that integrates nicely (in fact, we can see that s is just like y except for the minus sign and the integration constant).

In the next exercise, we use this result to derive the seemingly mysterious functions put forward in Exercise 2.

Exercise 3 Produce the curve y and arclength function s corresponding to the following functions:

a) $f(x) = x^2$

b) $f(x) = 1/x$

Using the Existence Theorem

Now we turn to a more "reasonable" function, but one whose differential arclength lack an elementary function antiderivative.

Arch of Sine Curve.—Seek the arclength of one arch of the sine curve—that is, the arclength from 0 to π. With $y(x) = \sin x$,

$$\mathtt{ds} = \sqrt{1 + y'^2} = \sqrt{1 + \cos^2 x}.$$

This differential arclength does not have an elementary antiderivative, so *Maple* returns an unevaluated integration expression:

```
int( ds(y,x), x );
          /
          |                 2 1/2
          |   (1 + cos(x) )       dx
          |
          /
```

Exercise 4 Show that the `ds(y,x)` above is continuous and hence that the Existence Theorem guarantees that any standard sequence of Riemann sums converges to the numerical value of the sought arclength. (You'll be asked to compute the arclength of one arch of the sine curve "In the Lab", see Exercise 7).

IN THE LAB

Exercise 5 Find out if *Maple* can produce an antiderivative for the differential arclengths $\sqrt{1 + y'^2}\, dx$ corresponding to the following functions:

a) $y = x^4$

b) $y = \sqrt{x}$

c) $y = x^{1/4}$

Exercise 6 Use *Maple* to compute antiderivatives for the arclengths in Exercise 2. Be sure to `simplify` the results and then print them out (see Exercise 11).

Exercise 7 Define y, a and b and use the code

```
evalf( Int( ds(y,x), x = a..b ) )
```

to compute the numerical value of the arclength of one arch of the sine function.

Exercise 8 Hand in a table of the arclengths of the sine function from 0 to b where $b = 0$, $\pi/8$, $2\pi/8$, ... π.

Exercise 9 Hand in a plot of the arclengths of the sine function from 0 to b as a function of b in the range 0 to π.

AFTER THE LAB

Exercise 10 Pick an $f(x)$ and use Equation 9 to construct a "doable" arclength problem. Solve your problem the straightforward way and check your answer against Equation 10.

Exercise 11 Reconcile the results of Exercises 2 and 6. Comment on the usefulness of *Maple* in finding these antiderivatives in a convenient form. Also comment on its usefulness in obtaining numerical values.

Exercise 12 Discuss the consistency of the results of Exercises 7, 8, and 9.

BEFORE THE LAB

Introduction

As you know, if a constant force F is applied over a distance d the work done is simply the product Fd. More frequently the force is changing due to any number of things (e.g., the terrain, the wind, a magnetic field, etc.). Moreover, often the force is not applied in a single direction, and this complicates things a bit. You will be exploring this important notion in a simple setting in which we assume:

a) The force is a function only of the horizontal variable x (i.e., $F = F(x)$).

b) The force is exerted directly along the curve of motion.

c) The curve of motion is a planar curve explicitly defined by $y = h(x)$ on some interval $[a, b]$.

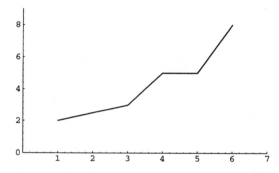

FIG. 14. The piecewise linear curve C_1.

Exercise 1 Without using calculus,

a) Calculate the length of the curve C_1, the piecewise linear curve connecting the points, $(1, 2)$, $(3, 3)$, $(4, 5)$, $(5, 5)$, and $(6, 8)$, shown in Figure 14.

b) Calculate the length of the curve C_2, defined by: $y = \sqrt{4 - x^2}$, for $-2 \le x \le 2$. **Hint**: Just use some geometry.

c) By approximating with a few inscribed chords, estimate the length of the curve C_3, defined by: $y = x^2 - 4$, for $0 \le x \le 2$.

Exercise 2 In this problem, k is a constant, $m(x)$ is the slope of the curves C_1, C_2, and C_3 defined in Exercise 1. Find $m(x)$ for each of these curves and then without doing any integrals,

a) Calculate the work done if the force $F = k \cdot m(x)$ is applied along the curve C_1.

b) Calculate the work done if the force $F = k$ is applied along the curve C_2.

c) Estimate the work done if the force $F = k \cdot m(x)$ is applied along the curve C_3.

Exercise 3 Now let C represent a general planar curve defined by $y = h(x)$ on the interval $[a, b]$ and let $F(x)$ be a force always exerted in the direction of C.

a) By considering the element of work done on a small interval, $[x, x + dx]$, carefully explain the following formula:

$$\text{Work} = \int_a^b F(x)\sqrt{1 + (h'(x))^2}\, dx$$

b) Use this formula to compute the work done by $F(x) = 3x$ along the curve C_3 defined above.

IN THE LAB

First set up the *Maple* statements to compute the length and work for the situation described in Exercise 3:

```
ArcLength := int( sqrt(1 + D(h)(x)^2), x = a..b );
Work := int( F(x) * sqrt(1 + D(h)(x)^2), x = a..b );
```

If `int` does not evaluate a given integral, follow it by an `evalf(")`. Note: if you are quite certain `int` will not do the job, avoid the attempt by using: `evalf( Int(...)  )`.

Exercise 4 Using *Maple*, compute the lengths of C_2 and C_3. Compare with your earlier results.

Exercise 5 Using *Maple*, compute the work associated with each of the following curves and forces:

a) Check your result for the second part of Exercise 3.

b) Let C be defined by
$$h(x) = \frac{\sin 3x}{1 + x^2}$$
on the interval $[0, \pi]$ and let $F(x) = 2x^2$. Compute both the length and work.

c) Compute the work for the last curve, but now with a force equal 3 times the slope of the curve.

AFTER THE LAB

Exercise 6 Considering the length of the curve and the magnitude of the force in the last part of Exercise 5, the work is rather small. Can you explain this?

Exercise 7 The second part of Exercise 3 was designed so that you could evaluate the work integral "by hand." Construct a *new* example of an h and F, so that you can similarly evaluate the work integral. **Challenge:** Construct a whole *family* of h's and F's that lead to easy work integrals.

BEFORE THE LAB

Introduction

An important problem in science and engineering is to find a functional relationship between two (or more) variables. For example, in some physical setting:

a) How are temperature and pressure related?

b) How are velocity and time related?

Often this question is addressed by examining a discrete set of data—e.g., one gathered by running several, or many, experiments. In this project you will be trying to establish an approximate linear (or quadratic) relationship from sets of data.

Exercise 1 Consider the following set of five x-values, followed by the y-values:
x-values: .2, 1.1, 2.0, 3.1, 3.9, 5.1
y-values: -2, 0, 1, 2.8, 5, 7

First carefully plot the five (x, y) points on a piece of graph paper. You will notice that the points almost, but not quite, lie on a line. Assuming that these points represented an approximation to a linear relation, attempt to find this relation by drawing a line on your graph that seems to best "fit" this data. Write down the equation of your line.

We now see how the problem of attempting to fit a line, or another curve, through a set of discrete data can be done using *Maple*. Let's put the data from Exercise 1 into some variables:

```
xdata := [.2, 1.1, 2.0, 3.1, 3.9, 5.1]:
ydata := [-2, 0, 1, 2.8, 5, 7]:
```

In *Maple* the xdata and ydata can be combined into a set of ordered pairs by the command:

```
data := zip( (x,y) -> [x,y], xdata, ydata);
    data := [[.2, -2], [1.1, 0], [2.0, 1], [3.1, 2.8], [3.9, 5], [5.1, 7]]
```

And then these ordered pairs can be plotted by:

```
gr1 := plot(data, style = POINT):
display(gr1);
```

as shown in Figure 15.

We have saved this graph in gr1 for later use. (The style = POINT option indicates that we want to see the points, rather than connecting lines.

Next follows the *Maple* commands that will find the "best line" (by some criteria) suggested by these points. First we call in the statistics package, then do a "linear regression" which finds this so-called best linear function.

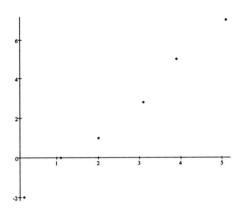

FIG. 15. Plot of the data.

```
with(stats):
cof := linregress(ydata, xdata);
                    cof := [-2.360934354, 1.815948449]
y := cof[1] + cof[2] * x;
                    y :=  - 2.360934354 + 1.815948449 x
```

Let's compute a graphic for the plot of this function on the interval $[.2, 5.1]$:

```
gr2 := plot(y, x = .2..5.1):
```

Now to display the two graphs together:

```
display( [gr1, gr2]);
```

as shown in Figure 16.

Note: This line fits this set of points "best" in the sense of "least squares" (in the errors). This will be a recurring subject in your career. This process is an example of "curve fitting" a discrete set of data, or, as said above, seeking a functional relation between x and y.

IN THE LAB

Exercise 2 The data below came from the Allegheny National Forest in Pennsylvania. The issue addressed was: can either the diameter or the height of a tree accurately predict the volume of wood in the tree? Using the data provided below, you are to see if there is a meaningful (linear) relation between:

    a) diameter and volume, or

    b) height and volume.

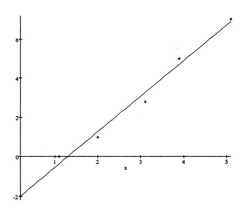

FIG. 16. Plot of the data with best fitting line.

Below, the "ddata" represents the diameter data for 11 different trees and "hdata" is the height data and "vdata" is the volume data for the same trees. Which of the two (diameter or height) would be the better estimator of volume? Roughly what volume would you expect from a tree with a diameter of 19 inches? How much error would you expect in this estimate? Why? Turn in two key graphs supporting your work. **Remark:** Think about whether **ddata**, etc., is analogous to **xdata** or to **ydata** in the above example.

```
ddata := [8.6, 10.7, 11.0, 11.4, 12.0, 13.3, 14.5, 16.0, 17.3, 18.0, 20.6]:
hdata := [65, 81, 75, 76, 75, 86, 74, 72, 81, 80, 87]:
vdata := [10.3, 18.8, 18.2, 21.4, 19.1, 27.4, 36.3, 38.3, 55.4, 51.5, 77.]:
```

Exercise 3 The following data came from velocity readings from a Porche 914 acceleration test. A plot of the points (velocity vs. time) shows that the relation is not linear. Find out if it is quadratic, i.e., see if $v = a + bt + ct^2$, or approximately so. Find the quadratic and show a key graph. First the time data, then the velocity data:

```
tdata := [0, 1.8, 2.8, 3.9, 5.1, 6.8, 8.6, 10.5, 13, 16.3, 19.7, 25.6]:
vdata := [0, 30, 40, 50, 60, 70, 80, 90, 100, 110, 120, 130]:
```

Note: this is no longer "linear regression," so we introduce a more general curve-fitting code. After "zipping" the data together, you will need these commands to first form the data 'matrix,' then solve for the coefficients a, b and c:

```
dataMat := array( [[t,v], op(data)]);
cof := regression(dataMat, v=a+b*t+c*t^2);
```

Exercise 4 In the previous problem, you will have seen that a quadratic fit, while much better than a linear fit, is still not terrific. By looking at the shape of the data, find other function(s) (i.e., other than or in addition to 1, t, t^2) that work better in the **regression** command.

AFTER THE LAB

Here we introduce the "method of least squares" which is the technique used by *Maple* in the above curve fitting. In our application of this important approximation technique, we have a set of n data points, (x_i, y_i) and want to find the line, $y = ax + b$, that best fits the n data points in the sense that the sum of the squares of the errors in this approximation is as small as possible. This is clearly a minimization problem and the function to be minimized is

$$F(a, b) = \sum_{i=1}^{n} [(ax_i + b) - y_i]^2.$$

Notice that despite all the x's and y's lying around, the *unknowns* are a and b—we want to find the a and b that minimize $F(a, b)$ for the *given* x_i's and y_i's.

Since we don't yet know how to minimize functions of more than one variable, we simplify the problem by assuming that we know the value of b, say, $b = 0$. Hence we seek the best line of the form $y = ax$. In this case $F = F(a)$ and we have a one-variable minimization problem that we *do* know how to handle.

Exercise 5

a) Assuming $b = 0$, show that in seeking the minimum of $F(a)$, one is led to the value:

$$a = \frac{\sum_{i=1}^{n} x_i y_i}{\sum_{i=1}^{n} x_i^2}.$$

b) Still assuming $b = 0$, find, by "hand," the least square line through the data: (1, 1), (2, 4), (3, 5), (4, 6).

c) On graph paper show the resulting line and the data points. Was the assumption that $b = 0$ supported by this particular data set?

Exercise 6 In Exercise 2, you may have been surprised that the attempt to find that a linear relation between diameter and volume of the trees was as successful as it was. What relation would you *expect* to find (think about the shape of a tree)?

Exercise 7 Again in reference to Exercise 2, speculate on why the data did not support finding a linear relation between height and volume of the trees.

BEFORE THE LAB

Introduction

We continue the theme of curve fitting a set of discrete data. However, in this project, the functional relationship in the data is a little more complex. Consequently, either the x-data or the y-data, or both, need to be "transformed." Typically, one has theory supporting a certain functional relationship between x and y, or such a relationship is suggested by the data itself. In this project, we study cases in which a logarithmic transformation of the data produces a linear relationship in the new (transformed) variables.

Exercise 1 Suppose we know, either from theory or empirically, that the power law relationship (or "model")

$$y = k\,x^m$$

holds between x and y. By taking the logarithm of this equation, show that one obtains the *linear* relationship,

$$Y = K + mX,$$

where $X = \ln x$, $Y = \ln y$, and $K = \ln k$ (so that we compute the model parameter k as $k = e^K$).

Exercise 2 Now consider the exponential model

$$y = c\,e^{ax}.$$

As in Exercise 1, obtain the linear equation

$$Y = C + aX$$

and identify X, Y and C in terms of the original model quantities. How would you compute the model parameters c and a from the fit to this linear equation?

Exercise 3 Now consider the "logistic" population model,

$$y = \frac{M}{1 + e^{c-ax}}\,, \qquad M > 0. \tag{11}$$

Show that if $y(0) < M$, then $y(x)$ is strictly increasing for $x > 0$, and also show that $\lim\limits_{x \to \infty} y = M$. To linearize the logistic relationship, temporarily think of $c - ax$ as the unknown and derive the equation $Y = c - ax$. What is Y in this equation?

Remark: In population models, M is the maximum sustainable population taking into account the restrictions in space, food, etc.

Illustrative Example

We revisit the tree data problem from the previous project on *Curve Fitting for Discrete Data Sets*. We start as before to get the data for the ordinary linear fit:

```
ddata := [8.6, 10.7, 11.0, 11.4, 12.0, 13.3, 14.5, 16.0, 17.3, 18.0, 20.6]:
vdata := [10.3, 18.8, 18.2, 21.4, 19.1, 27.4, 36.3, 38.3, 55.4, 51.5, 77.]:
data := zip( (x,y) -> [x,y], ddata, vdata);
```

Recall the **zip** command merges the two set of data together for plotting purposes. For later comparison, we get the old linear fit (see Figure 17):

```
with(stats):
cof := linregress( vdata, ddata);
linearFit := cof[1] + cof[2]*x;
                    linearFit :=  - 41.13128168 + 5.385554747 x
```

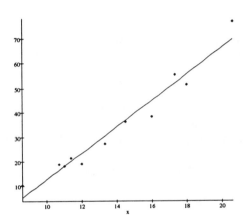

FIG. 17. Linear fit to the tree data.

As in Exercise 1, we compute the logarithm of the data. Here is *Maple*'s way of taking the log of two lists of data:

```
lnDdata := map( x -> ln(x), ddata):
lnVdata := map( x -> ln(x), vdata):
```

Once again one would need to **zip** the new data sets together if plotting the points. Here is the code to get a fit to the logs of the data:

```
cof := linregress( lnVdata, lnDdata);
linearFit := cof[1] + cof[2]*d;
                    linearFit :=  - 2.468542415 + 2.241055600 d
```

Finally, copy over the m-value and use the K-value to get k (cf. Exercise 1) and thus obtain the new power law fit:

```
m := cof[2];
k := exp(-cof[1]);
```

A plot of the power law fit, $k\,x^m$, vs. the original data is shown in Figure 18. The improvement over the linear fit is substantial.

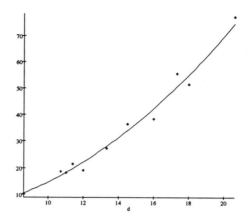

FIG. 18. Power law fit to the tree data.

IN THE LAB

Exercise 4 Below is data relating the period of revolution T (in days) of the 6 inner planets and their semi-major axes a (in 10^6 km). Find k and m in Kepler's empirical power law relationship, $T = ka^m$. Hand in one graph showing the effectiveness of your fit. Newton was later able to justify the theoretical value $m = 3/2$ by assuming an inverse square force law for gravitational attraction—how well does the empirical data bear out Newton's theory?

```
adata := [58, 108, 149, 228, 778, 1426]:
Tdata := [87.97, 224.70, 365.26, 686.98, 4332.59, 10759.20]:
```

Exercise 5 Below is data pertaining to the 1988 Olympic weight-lifting competition. The **xdata** is the "class data" giving seven weight classes from Featherweight to Heavyweight-2 (given in kg). The **wdata** is the combined weight lifted by the winners in each class.

```
xdata := [56, 67.5, 75, 82.5, 90, 100, 110]:
wdata := [292.5, 340, 375, 377.5, 412.5, 425, 455]:
```

a) Seek to fit this data in the form $w = kx^m$ as in Exercise 4. Find k and m. (Note: theoretically, one should get $m = 2/3$. More on this later).

b) Add to the data one weight class that we had left out: N. Suleymanoglu of Turkey won the 60 kg class with a combined weight of 342.5 kg. Repeat the work of the previous part with the complete set of weight classes.

c) Comment on what the extra entry in the previous part did to your experiment. And explain why Suleymanoglu was referred to as "the strongest man in the world."

Exercise 6 Below is data for the U.S. population from the years 1800 to 1990. The years are in the **xdata** array (year 1800 is denoted by 0, etc.) and the population, in millions, is given in the **ydata** array. First show that the popular exponential growth law, $y = c \exp ax$

does not work very well. Illustrate this failure with a graph. Then, as in Exercise 3, seek the relationship $y = M/(1+\exp(c-ax))$ and linearize to get $Y = c - ax$. Here you will need to make an "educated guess" at M, the maximum population attainable. For example, is $M = 300$ (representing 300,000,000) a good choice or will a smaller or larger M make a better model? You are to investigate this, trying to find an M that makes the model work well. (There is no perfect M). In particular, find an M that you think will work well in estimating the population in the year 2050 ($x = 2050 - 1800 = 250$). What is your resulting estimate?

```
xdata := [0, 20, 40, 60, 80, 100, 120, 140, 160, 170, 180, 190]:
ydata := [5.3, 9.638, 17.07, 31.44, 50.19, 76.2, 106.0, 132.2,
            179.3, 203.3, 226.5, 248.7]:
```

Useful commands for Exponential Fit:

```
lnYdata := map( y -> ln(y), ydata):
expdata := zip( (x,y) -> [x,y], xdata, lnYdata):
```

Useful commands for Logistic Fit:

```
Ydata := map( y -> ln(M/y - 1.0), ydata);
logisticdata := zip( (x,y) -> [x,y], xdata, Ydata):
```

AFTER THE LAB

Exercise 7 While doing Exercise 6, you should have tried several values of M. Explain why, in using your model to predict the population at year 2050, you selected the value of M that you did.

Exercise 8 In connection with Exercise 5:

   a) Explain why the weight w that a person can lift should behave like $w = k\, x^{2/3}$, where x is the weight of the person.

   b) Use this result to describe a fair method for doing away with weight "classes" altogether (often needed in small meets).

Hint: Assume that strength is roughly proportional to the cross-sectional area of muscle and that body weight is roughly proportional to body volume for persons of comparable fitness.

BEFORE THE LAB

Introduction

Simple harmonic motion is a fundamental model in science and engineering. The study of the simple differential equation, $y''(t) + k^2 y(t) = 0$, is the first step in the mathematical analysis of the vibration of physical systems as well as in the analysis of the propagation of sound, light and electromagnetic waves. The elementary solutions to this differential equation, namely $\sin kt$ and $\cos kt$, are also used in modeling a host of other phenomena that have cyclical or near-cyclical behavior.

This project and the later one on *Fourier Frequency Decomposition* will show you how these important functions can combine in some interesting ways and, conversely, how "signals" (or functions) can be decomposed naturally into their frequency components.

Exercise 1 Suppose the displacement $y(t)$ of a particle satisfies: $y'' + 9y = 0$ and at time $t = 0$ has a 4 cm displacement and a 6 cm/sec velocity. Solve for $y(t)$ in both the forms $a \cos kt + b \sin kt$ and $A \cos k(t - \alpha)$.

Exercise 2 For the function $f(x) = \cos x - \sin x$:

a) Write f in the form $A \cos k(x - \alpha)$. **Note:** α turns out to be *negative*.

b) Find its range.

c) Find the largest domain including $x = 0$ for which f is decreasing.

d) Find the inverse of f and its domain and range.

IN THE LAB

Exercise 3 Using *Maple* confirm that your work in Exercise 2 is correct. In particular, plot both f and its inverse (restricted to the proper domains) on the same graph, thus demonstrating the theoretically predicted symmetry about the line $y = x$.

Exercise 4 This exercise gives the background for the study of sound and other traveling waves. Consider the function:

$$y = \sin Ax + \sin Bx,$$

where A and B are constants. We will study the effect when A and B change relative to each other. Note that A and B (the "circular frequencies") are measured in radian/sec. Set A to a number of your choice, say between 4 and 10; this is your fundamental frequency and is to remain fixed.

a) First set $B = A$ and describe the predictable result for the frequency and amplitude of y.

b) Now change B by a small amount, say 5 to 10%—*be sure to plot over a large enough time interval to see what is happening*. The results may surprise you—so carefully describe what you see regarding the amplitude and frequencies that you observe. For example, do you see the original frequency A, or has it been altered? Do you see a much lower frequency?

AFTER THE LAB

Exercise 5 You are to explain the two new frequencies that you observed in the second part of Exercise 4. **Hint:** The trigonometric identity,

$$\sin C \cos D = \frac{1}{2}\left(\sin(C + D) + \sin(C - D)\right),$$

will help. Compute the two new frequencies and point them out on your graph. In your experiment, what is the period of the "beats" (the beats occur at the times of maximum amplitude)?

BEFORE THE LAB

Introduction

The project on *Simple Harmonic Motion* illustrated "harmonic motion" in which a mass (or a sound wave, etc.) vibrates at certain frequencies. These vibrations were represented by sine and cosine functions. Harmonic motion is one of the fundamental phenomena of nature, and is widely used by scientists and engineers in modeling these phenomena. In these studies, a key step is analyzing the "frequency content" of the vibrating object.

For example, the function

$$y = \sin 11x \cos x$$

can, by a trigonometric identity, be written:

$$y = \tfrac{1}{2} \sin 10x + \tfrac{1}{2} \sin 12x.$$

This latter form is often preferred because it reveals that function y is the sum of the two components with circular frequencies 10 and 12 rad/sec (assuming x is time in seconds). We say that we have "decomposed" the original function into its frequency components. Each component might represent a frequency of oscillation, or a color, or chemical compound or whatever the application calls for. The fruitful discovery that many functions can be so decomposed was discovered by J.B.J. Fourier in 1807. In particular, Fourier argued that most functions of interest can be written in terms of their frequencies; i.e. in the form:

$$\begin{aligned} f(x) \;=\; & b_1 \sin x + b_2 \sin 2x + \ldots + b_n \sin nx \\ & + a_0 + a_1 \cos x + a_2 \cos 2x + \ldots + a_n \cos nx \end{aligned} \tag{12}$$

where n is an integer. In principle, n can be large or even infinite, but for our discussion, we will assume that n is small. For simplicity, in our examples for this project, we will also assume that $f(x)$ is "odd" in which case all the cosine terms will be zero (as you will show in Exercise 1). The a_k and b_k in Equation 12 are called *Fourier coefficients* and in the first exercise, you will find out how to compute them for the case just described. We say that two functions f and g are *orthogonal* on $[-\pi, \pi]$, if

$$\int_{-\pi}^{\pi} f(x)g(x)\,dx = 0.$$

(The word "orthogonal" is used here since this definition really *is* a (vast) generalization of the notion of perpendicular lines in geometry. However, explaining these subtle connections would take us far afield of the subject examined in this project—this is a major topic in undergraduate Linear Algebra courses.)

Exercise 1 Assume that $f(x)$ is an odd function (use the index of your text, if you don't know the definition of "odd function").

    a) If g is even, show that f and g are orthogonal on $[-\pi, \pi]$ (or, for that matter, on any interval $[-c, c]$).

    b) Show that $\sin kx$ and $\cos px$ are orthogonal on $[-\pi, \pi]$ for any positive integers k and p.

c) For k and p *distinct* positive integers, show that $\sin kx$ and $\sin px$ are also orthogonal on $[-\pi, \pi]$. (You will need a trigonometric identity here).

d) Suppose a given f is to be decomposed as in (12) above. To find a coefficient b_k, for $1 \le k \le n$, multiply Equation 12 through by $\sin kx$ and integrate over $[-\pi, \pi]$. Use the orthogonality shown in parts (b) and (c) to show that

$$b_k = \frac{1}{\pi} \int_{-\pi}^{\pi} f(x) \sin kx \, dx.$$

e) Now similarly seek the a_k in (12), showing that all the a_k are zero.

Exercise 2 Suppose you wish to expand the specific function, $f(x) = x$, in a Fourier series as in (12).

a) Show that $D(\sin kx - kx \cos kx) = k^2 x \sin kx$.

b) Using (a) and Exercise 1, show that

$$b_k = -2 \frac{(-1)^k}{k}.$$

c) Explain why the $a_k = 0$.

We now show how *Maple* can be used to decompose a function into its frequencies or, stated differently, to find its Fourier expansion. Again we'll consider only the case when $f(x)$ is odd, hence only the sine terms in (12) above are needed.

We use only a finite number of sine terms, so the expansion will not be exact, but only an approximation to the given $f(x)$.

Note: Since all the functions we consider will be odd, in computing the b_k coefficients, we integrate only over $[0, \pi]$ and then multiply the result by 2. (You will be asked to justify this later).

We start with a cubic polynomial and a small n to see what happens.

```
f := x -> x*(Pi - x)*(Pi + x):
nmax := 2: npi := evalf(Pi):
b := seq( 2/npi * int( f(x)*sin(k*x), x=0..npi), k=1..nmax);
                b := 12.00000000, -1.500000004
fn := x -> sum( b[n]*sin(n*x), n = 1..nmax):
fn(x);
                12.00000000 sin(x) - 1.500000004 sin(2 x)
plot({f, fn}, 0..Pi);
```

See Figure 19 for the result of the above **plot** command.

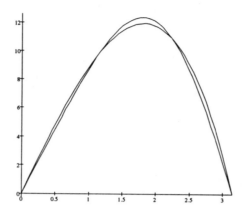

FIG. 19. The function $x(\pi - x)(\pi + x)$ and its two term Fourier approximation.

IN THE LAB

Exercise 3 For the above cubic, make n larger, until the approximation f_n agrees well with f graphically. Hand in a graph with appropriate comments.

Exercise 4 For $f(x) = x^2 \cot(x/2)$, start with a small n, and increase as necessary, compute a graphically accurate Fourier sine approximation f_n on interval $[0, \pi]$. Print out one graph showing both f and f_n.

Now consider the differential equation (or DE):

$$y'' + 8y = x^2 \cot(x/2) \tag{13}$$

The right side of Equation 13 is called the *forcing term*, because in mechanical applications it often represents an external force applied to the mass whose motion is described by y. Since this is a hard DE, we seek an approximate solution. From Exercise 4, you already have a good approximation to the difficult function $x^2 \cot(x/2)$ in the form:

$$b_1 \sin x + b_2 \sin 2x + \ldots + b_n \sin nx, \tag{14}$$

and you know an appropriate value for n. The idea is to seek a solution to the DE in (13) in the form:

$$y = c_1 \sin x + c_2 \sin 2x + \ldots + c_n \sin nx. \tag{15}$$

In Exercise 6 (after the lab) you will show that if (15) makes sense, then the c's and b's are related as follows:

$$c_k = \frac{b_k}{8 - k^2}, \qquad k = 1, 2, \ldots, n. \tag{16}$$

Exercise 5 As just discussed, find the appropriate approximation (e.g. call it y_n) to the solution to (13) and plot it. If you know how to solve this DE numerically, do so. Then compare at several points and attempt to determine the accuracy of your series solution over the interval $[0, \pi]$.

Fourier Frequency Decomposition

AFTER THE LAB

Exercise 6

a) In Exercise 1d you found the formula for the b_k. Explain why if f is odd, it is only necessary to integrate over $[0, \pi]$ as in the above *Maple* code.

b) Explain the relationship in (16) by substituting (15) into the DE using (14) for the right side of the DE.

c) Your approximate solution to (13) found in doing Exercise 5 has the n frequencies given in (14) and (15). However, from your earlier work with equations like $y''+8y = 0$, you might expect another frequency not accounted for. What is this frequency?

Exercise 7 (Challenge). How do you tell if the "missing" frequency in the previous exercise is needed? Answer: Commonly in solving DEs like the one in (13), one has initial conditions that also need satisfying, that is, the values $y(0)$ and $y'(0)$ are specified in addition to the DE. If the approximate solution found in Exercise 5 does not satisfy these conditions, you need another piece. Think about how you solved "homogeneous" equations like $y''+8y = 0$ and see if you can get a solution for arbitrary initial conditions, $y(0) = a$ and $y'(0) = b$, where a and b are constants.

BEFORE THE LAB

In this project we examine the mechanics of a rotating wheel which drives an arm. In practice, the arm could be connected to a piston or to a cutting blade. Mathematically, some interesting behavior takes place, including some rather unusual periodic functions and surprising results as the length of the arm approaches the radius of the wheel. Examine

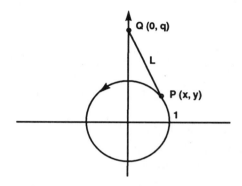

FIG. 20. Sketch for the piston problem.

Figure 20. The wheel is rotating in the counterclockwise direction at 2 radians/sec. The lower end of the arm at the point $P(x, y)$ is attached to the circle and rotates with it. The upper end of the arm at the point $Q(0, q)$ is attached to the vertical axis and can only move up and down on this axis.

Exercise 1 In connection with the Figure:

a) How fast (linear velocity) is point P moving?

b) Obtain this expression for q:

$$q(t) = \sin 2t + \sqrt{L^2 - \cos^2 2t}.$$

 Hint: What are the coordinates of P?

c) Compute $q'(t)$.

Exercise 2 You are not expected to know the "right" answer to the following questions— just use your imagination to gain an initial orientation to the situation. You will be able to later check your conjectures against what actually takes place.

a) Try to visualize the movement of point Q, as a function of t, as the wheel turns. First for L about the size shown, then for L slightly larger than the radius (1). Make rough sketches by hand.

b) Would you expect velocity of Q, $q'(t)$, to ever exceed that of P? Why?

c) For what location of P would you expect $q'(t)$ to be largest?

IN THE LAB

Exercise 3

a) Compute $q'(t)$ and check against your hand work done earlier.

b) For L considerably larger than the radius 1 (e.g. 2 or 3), graph $q(t)$ and $q'(t)$. Any surprises here?

c) Now experiment with L only slightly larger than 1, obtaining graphs of q and q' that are rather strange (with nearly flat sections). Comment on what you see, especially try to explain these flat sections.

AFTER THE LAB

Exercise 4

a) Revisit your conjectures of Exercise 2 and compare with the actual results you obtained here.

b) A theoretical physicist is interested in what happens as $L \to 1$. In particular, she is interested in the maximum of q' as $L \to 1$. Do you think, from your experiments, that this mysterious maximum value is finite or infinite? If finite, what do you think it is?

c) Briefly, if one were to explore, mathematically, the question raised in b), how would you pursue it?

BEFORE THE LAB

This project reviews the methods that *Maple* affords to do integration. Because it seems likely that in the next few years engineers and scientists will replace much of the older techniques of integration with tools like *Maple*, this project contains a significant amount of new *Maple* content aimed towards building expertise in doing integration with *Maple*.

Overview: Using *Maple* to do Integration

The Indefinite Integral or Antiderivative.—The indefinite integral is just another name for the antiderivative. It is usually denoted by

$$\int f(x)\, dx,$$

but occasionally is written with notations such as $D^{-1}f(x)$. The *Maple* syntax is

```
int(f(x), x);
```

Typical usages are

```
int(x^3, x);
```

$$
\begin{array}{c}
4 \\
1/4\ x
\end{array}
$$

and

```
f := x -> sin(x):
int(f(x), x);
```

$$- \cos(x)$$

Notice that *Maple* does not show the arbitrary constant of integration—you have to tack that on yourself.

The Definite Integral when an Elementary Antiderivative Exists.—We've seen this several times before. Here are some typical examples:

```
int(x^5, x = 10..20);
```

$$10500000$$

```
int(x^5, x = a..b);
```

$$
\begin{array}{cc}
6 & 6 \\
1/6\ b & -\ 1/6\ a
\end{array}
$$

(Make sure you understand the above output.) Now consider the following three variations on the purely numerical problem $\int_0^{1/2} \arcsin x\, dx$:

```
int( arcsin(x), x = 0..1/2);
                                                        1/2
                             1/12 Pi + 1/2 3      - 1
evalf( int( arcsin(x), x = 0..1/2) );
                                      .127824792


evalf( Int( arcsin(x), x = 0..1/2) );
                             .127824792
```

In the first case, the *exact answer* is obtained. In the second case, the exact answer is obtained and then converted to an approximate decimal answer. In the final example, a numerical method is evoked immediately. The latter two methods which each give numerical answers exhibit some trade-offs. For example, if *Maple* does not "know" the antiderivative, `evalf( int(...))` will be slower. However, if there are numerical difficulties, it may be more accurate. The non-expert should just be aware that if one command doesn't work well in a given situation, the other should be tried.

For a last example of definite integration with elementary antiderivative, consider the "Fundamental Theorem" type integral:

$$\arcsin(x) = \int_0^x \frac{dt}{\sqrt{1 - t^2}}.$$

It is easy to check that this integral is, indeed, $\arcsin x$. A natural first coding is (we suppress the actual plot, since we just want to make a point about how long the execution takes):

```
MyArcSin := x -> int( 1/sqrt(1 - t^2), t = 0..x):
plot(MyArcSin, -0.9..0.9);
```

Notice how long this takes to plot. This is slow because for each plot-point the integration is done; and this is characteristic of the formal function (like MyArcSin above) since evaluation is postponed until the function is used. One way to speed things up is to assure that the integration is done up front. For example,

```
y := int( 1/sqrt(1 - t^2), t = 0..x);
                              y := arcsin(x)
plot(y, x = -0.9..0.9);
```

Now it is clear that the integration is performed immediately upon defining y; and the plotting will be much faster.

The Definite Integral when there is no Elementary Antiderivative.—
The function definition

$$\text{superarcsin}(x) = \int_0^x \frac{dt}{\sqrt{1-t^4}}$$

is analogous to the representation of the arcsin given in the previous section ("superarcsine" is just a made-up name). But here there is no elementary antiderivative:

```
SuperArcSin := x -> int( 1/sqrt(1 - t^4), t = 0..x);
                                     x
                                    /
                                    |       1
            SuperArcSin := x ->     |  ------------ dt
                                    |        4
                                    /    sqrt(1 - t )
                                   0
```

The above output shows that *Maple* does not "know" about this integral (but some version of *Maple* may express the answer in terms of the non-elementary 'hypergeometric' function). Since *Maple* did not succeed in finding an antiderivative there is no way to avoid the slow plotting of SuperArcSin. Nevertheless its plot is shown in Figure 21.

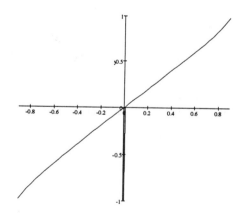

FIG. 21. The "superarcsine" function.

As another example of a function for which we have no antiderivative we define a new function called "supersine." Again, "supersine" is just a made-up name and, in particular, is *not* the inverse function to the "superarcsine" function invented earlier. Once again *Maple* was not successful in the integration and plotting was quite slow.

```
SuperSin := x -> int( sin(sin(t)), t = 0..x);
plot( SuperSin, -3*Pi..3*Pi);
```

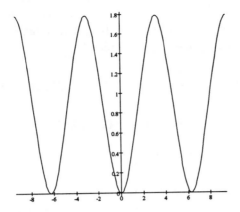

FIG. 22. The "supersine" function.

Exercise 1 Figure 22 shows that the supersine function is never negative. Explain mathematically why this is so. **Hints:** (1) To get some insight, consider the easier, but analogous, integral $\int_0^x \sin t \, dt$. (2) To polish off this problem, think about the area interpretation of the definite integral.

Exercise 2 Consider each of the following indefinite integrals. Try to evaluate them by hand or by using tables or by using calculators or by any other means you have at your disposal outside the lab. *Neatly* write up your best attempts on each problem, explaining *in detail* your methodology. These explanations may well include statements like "I had it essentially right, but after running *Maple* (see Exercise 4), I discovered several mistakes ..."

a) $\displaystyle\int x^{11} \sqrt[3]{x^4 + 1} \, dx$

b) $\displaystyle\int \frac{dx}{\sqrt{e^{4x} - 1}}$

c) $\displaystyle\int x^5 e^{ax} \, dx$

d) $\displaystyle\int x^3 \cos ax \, dx$

e) $\displaystyle\int x^3 \sin ax \, dx$

IN THE LAB

Exercise 3 The superarcsine function appears similar to the ordinary arcsine function. Hand in a graph that proves they aren't really the same. Describe how they differ.

Exercise 4 Try *Maple* on the integrals in Exercise 2. Be aware that *Maple* may write the result differently than your hand calculations. Plotting the two forms on the same graph will reveal if they are the same (or differ only by an integration constant). Use your lab results to augment the report required by Exercise 2.

Exercise 5 Solve the equation, supersine(x) = 1, to at least 8 significant figures.

AFTER THE LAB

Nothing to do this time.

The Probability Integral (optional)

BEFORE THE LAB

Integration of Even and Odd Functions

Exercise 1 Evaluate the sequence of integrals, $I_n = \int_{-3}^{3} x^n \, dx$ for $n = 1, 3, 5$.

Exercise 2 Suppose that f is an *odd* function, that is: $f(-x) = -f(x)$. Prove that $\int_{-a}^{a} f(x) \, dx = 0$. **Hint**: Break the integral into two parts around $x = 0$ and use the *substitution $u = -x$ on the integral from $-a$ to 0.*

Exercise 3 Suppose that f is an *even* function, that is: $f(-x) = f(x)$. Prove that $\int_{-a}^{a} f(x) \, dx = 2 \int_{0}^{a} f(x) \, dx$. This has practical use because evaluating at $x = 0$ is less error prone than evaluating at a negative number.

The Method of Substitution

Exercise 4 Let $I = \int \sin x \cos x \, dx$, use the substitution $u = \sin x$ to show that $I = \frac{1}{2} \sin^2 x + C$. Then use the substitution $u = \cos x$ to show that $I = -\frac{1}{2} \cos^2 x + C$. Is anything wrong here? Explain.

The function

$$n(x) = \frac{1}{\sqrt{2\pi\sigma^2}} e^{-(x-\mu)^2/2\sigma^2}$$

is the famous and important probability density for the *normal* distribution with mean value μ and standard deviation σ, see Figure 23.

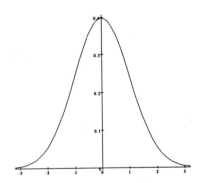

FIG. 23. The normal probability density.

If the "random variable" x follows the normal distribution, then the probability that x lies in the interval $[a, b]$ is defined as

$$P = \int_{a}^{b} n(x) \, dx.$$

It is often of interest to know the probability of lying within 1, 2, or in general, k, standard deviations of the mean. This is given by the integral

$$P(k) = \int_{\mu-k\sigma}^{\mu+k\sigma} n(x) \, dx. \tag{17}$$

Indeed, everyone taking an elementary statistics course memorizes the following:

Rule of Thumb for the Normal Distribution: Two-thirds of the observed values lie within one standard deviation from the mean, 95% lie within two standard deviations, and 99% lie within three standard deviations.

Among other things, we'll be checking out this Rule of Thumb in this project. One famous (or infamous?) application is "grading on the curve": those students scoring within one standard deviation of the mean earn C's, while those scoring between one and two standard deviations above the mean earn B's and those scoring over two standard deviations above the mean earn A's and similarly at the low end for the D's and F's.

Exercise 5 Perform the substitution $z = (x - \mu)/\sigma$ to show that

$$P(k) = \frac{1}{\sqrt{2\pi}} \int_{-k}^{k} e^{-z^2/2} \, dz. \tag{18}$$

Thus, $P(k)$ does not depend on μ or σ. This is important if you want to make tables for P—you need only one table instead of infinitely many. Notice also that this exercise is an application of substitution in integrals *not* aimed at getting an antiderivative.

As $k \to \infty$, we include all the probability, so you'd expect $\lim_{k\to\infty} P(k) = 1$ and *Maple* agrees (in *Maple* ∞ is written `infinity`):

```
p := x -> exp(-x^2/2) / sqrt(2*Pi);
int( p(x), x = -infinity.. infinity) ;
                                  1
```

What is the probability that a value lies within one standard deviation? Now *Maple* says:

```
 int( p(x), x = -1.. 1);
                             1/2
             erf(1/2 2    )
```

erf? Well, indeed, erf is the name of a special function (the "error" function) closely associated with the normal distribution and this is the best *Maple* can do because the normal distribution does not have an elementary function antiderivative:

```
int( p(x), x);
                             1/2
         1/2 erf(1/2 2    x)
```

However, *numerical* values are no problem, since *Maple* has methods for evaluating the erf function:

```
onesigma := evalf( int( p(x), x = -1.. 1) );
                   onesigma := .6826894920
```

Thus 68.3% of the probability is within one standard deviation of the mean. This is a more precise estimate than that given by the rule of thumb value (2/3 or 66.7%).

The Probability Integral (optional)

Exercise 6 The *standard* normal distribution is the normal distribution with mean $\mu = 0$ and standard deviation $\sigma = 1$. Show that the probability density function for the standard normal is $p(x) = \frac{1}{\sqrt{2\pi}}e^{-x^2/2}$. Then show that the simplified form in Equation 18 of $P(k)$ is just the original form of Equation 17 with n replaced by p. (This explains why only the *standard* normal is tabled in handbooks.)

Method of Substitution with *Maple*

We know that the differential is a formal guide to change of variables in an integral: $f(x)\,dx$ becomes $f(g(u))\,g'(u)\,du$ under the one-to-one change of variables $x = g(u)$. We can follow that insight closely in *Maple*:

Illustrative Example: Simplify the integral of

$$\frac{1}{1 + e^{2x}}$$

by making the change of variables $u = e^x$.

Answer: To apply the above theory, we need the substitution in the form $x = g(u)$, that is, $x = \ln u$:

```
f := x -> 1 / (1 + exp(2*x)):
g := u -> ln(u):
integrand := u -> f(g(u))*D(g)(u):
integrand(u);
```

$$\frac{1}{(1 + u^2)\, u}$$

Now we perform the integration in u and then return to the x-variable by using the 'inverse' function $u = e^x$ (which is really the way we humans start off).

```
answeru := int( integrand(u), u);
ginverse := x -> exp(x):
answerx := subs( u = ginverse(x), answeru);
```

$$answeru := \ln(u) - 1/2\,\ln(1 + u^2)$$

$$answerx := \ln(\exp(x)) - 1/2\,\ln(1 + \exp(x)^2)$$

```
simplify(");
```

$$x - 1/2\,\ln(1 + \exp(2\,x))$$

In this case *Maple* can do the original integral, which provides a check on our work:

```
simplify( int(f(x), x) );
```

$$x - 1/2\,\ln(1 + \exp(2\,x))$$

Now we show how to handle limits. Suppose we have the same integrand to be integrated from 0 to 1. Now we do not have to return to x, but we still need the 'inverse' function, this time for the limits! Starting from `integrand(u)` again:

```
a := 0:  b := 1:
A := ginverse(a):  B := ginverse(b):
int(integrand(u), u = A..B);
```

$$1 - 1/2 \ln(1 + \exp(1)^2) + 1/2 \ln(2)$$

IN THE LAB

Exercise 7 plot $\frac{1}{2}\sin^2 x$ and $-\frac{1}{2}\cos^2 x$ on the same graph and relate the result to Exercise 4. Note: In *Maple*, $\sin^2 x = $ sin(x)^2.

Exercise 8 Using the definition in Equation 18, show that $P(k) = \sqrt{2/\pi} \int_0^k e^{-z^2/2}\,dz$. Translate this into a *Maple* function definition and numerically evaluate $P(1)$, $P(2)$ and $P(3)$. Explain what these results mean in terms of probabilities and compare them to the rule of thumb given above. Explain how one can use Figure 23 to get a rough check that all is well.

Exercise 9 For testing statistical hypotheses, it is also important to know the value of k (not usually an integer) such that $P(k)$ equals a given value v. The values $v = .95$ ("significant") and $v = .99$ ("highly significant") are especially popular. To six significant figures, find the k values corresponding to the three v values, .90, .95 and .99. **Hint:** Set up Newton's method, but watch out, the letters are non-standard! In particular, k is the independent variable, *not* a constant. Also, be careful about the starting value—a bad one could cause divergence.

Exercise 10 Verify that *Maple* cannot evaluate

$$\int_0^{\pi/2} \cos(z \cos x)\,dx.$$

As in the Illustrative Example above, apply the 'obvious' substitution $u = \cos x$ and then evaluate the resulting u-integral as a function of z, say $h(z)$. Find out how many zeroes the function $h(z)$ has in the interval $-12 < z < 12$.

AFTER THE LAB

Nothing this time.

BEFORE THE LAB

Exercise 1 Do the following "by hand."

a) Find the Taylor polynomial of degree three, $P_3(x)$, by expanding the function $f(x) = e^x \sin x$ about $x = 0$. **A check**: You should get $f^{(iv)}(x) = -4e^x \sin x$ for the fourth derivative.

b) Using the remainder term in Taylor's theorem, determine an upper bound for the absolute value of the maximum error in $P_3(x)$ on the interval $[-1, 1]$.

Exercise 2 The following idea of building on known expansions is often used to get an approximation for a function. Consider, for example, the function $g(x) = \sin(\sin x)$.

a) Explain why $\sin x \approx x - \frac{1}{6} x^3$ is a decent approximation for "small" x.

b) Suppose that the above small x approximation is suitable for your purposes. You seek a similar approximation for $g(x)$. Let $u = x - \frac{1}{6} x^3$ (your approximation to $\sin x$), and now similarly approximate $\sin u$. Thus, by expanding in powers of x obtain $g(x) \approx x - \frac{1}{3} x^3$. (You will see how well this works later in this project).

Taylor Series in *Maple*

`taylor( f(x), x = a, n)` is the *Maple* command to compute the Taylor polynomial P_n of degree n about $x = a$ for a given function f. (The more general command `series` also works, but we will stick with `taylor`). Here is how it works:

```
f := x -> x * cos(2*x):
a := 0:  n := 6:
ourSeries := taylor( f(x), x = 0, n);
                3        5        7
           x - 2 x  + 2/3 x  + O(x )
```

Here we asked for terms through x^5 by setting $n = 6$. In this case, however, *Maple* knew that the x^6 term was zero (since $\sin 0 = 0$); hence the $O(x)^7$ at the end of the output designates that the error term is of "order" 7. In *Maple*, expressions containing the O symbol are treated as approximate and, hence indefinite, quantities. Before you can plot, or otherwise use the Taylor polynomial in ordinary calculations, you must convert to the corresponding ordinary expression by using the `convert` command to create the desired polynomial, as follows. Recall that we defined the above series in a variable called 'ourSeries,' so:

```
Pn := convert(ourSeries, polynom);
                   3        5
        Pn := x - 2 x  + 2/3 x
```

Let's see how good our approximation is on the interval $[-1, 1]$. The command

```
plot( {f(x), Pn}, x = -1..1);
```

produces Figure 24 showing that the approximation is close, although it is breaking down at the ends of the interval. To better see the error, plot the difference $f(x) - Pn$; see Figure 25.

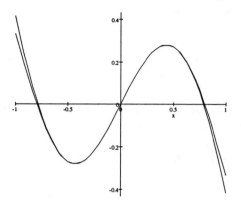

FIG. 24. $x \cos 2x$ and its 5th degree Taylor polynomial at $x = 0$.

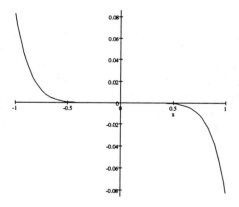

FIG. 25. The difference between $x \cos 2x$ and its 5th degree Taylor polynomial at $x = 0$.

IN THE LAB

Exercise 3 For the function $f(x) = x \cos 2x$ discussed above:

a) Gradually increase n until you have a good graphical approximation. Show your final graph.

b) Suppose you wish to approximate $f(x)$ on the interval $[-1, 1]$ by an nth degree polynomial $P_n(x)$ to within an error of 0.005. By plotting the difference $f(x) - P_n(x)$, experimentally find the smallest n that is adequate. Show your final graph of $f(x) - P_n(x)$.

Exercise 4 Following up on Exercises 1 and 2:

a) Use the `taylor` command to find the $P_3(x)$ in Exercise 1. Graph the error $f(x) - P_3(x)$ and comment on how the actual error compares with your upper bound in Exercise 1.

b) Use the `taylor` command to find a Taylor polynomial approximating $g(x) = \sin(\sin x)$, thus checking your earlier work in Exercise 2.

c) Plot $g(x)$ and your cubic approximation together on each of the intervals $[-0.5, 0.5]$ and $[-1, 1]$. Comment on the quality of your approximation in each case.

Exercise 5

a) Obtain series approximations to $\sin(\tan x)$ and $\tan(\sin x)$ about $x = 0$ and determine the first power of x for which the series differ.

b) Since these two series look very much alike, you may think they represent essentially the same functions. Hand in plots of $\sin(\tan x)$ and $\tan(\sin x)$ on intervals $[0, \pi/4]$ and $[0, \pi]$ and comment.

c) One of the functions behaves wildly around $\pi/2$. Which one? Explain this strange behavior.

Solving a Differential Equation via Taylor Approximations

Here, we use power series to solve a differential equation. On the interval $[-1, 1]$, we want to solve the differential equation,

$$y'' = g(x) = e^{-x^2} \sin x,$$

with initial conditions, $y(0) = 0.3$ and $y'(0) = -0.1$. This is reminiscent of the falling body problem, except now the right side, $g(x)$, is *not* the earth's gravitational constant. Thus, we will have trouble integrating y'' to get y', etc. So we seek an approximation to the solution, y, as in the next exercise.

Exercise 6

a) Integrate $g(x)$ using *Maple*. What happens?

b) Approximate $g(x)$ by a Taylor polynomial $P_n(x)$ for a large enough n so that the approximation looks good on $[-1, 1]$.

c) Now integrate $P_n(x)$ twice with *Maple*, applying the initial conditions, thus getting a (hopefully) suitable approximation to our desired solution. Provide a graph of the resulting y.

AFTER THE LAB

Nothing this time.

Introduction

We've seen several functions that are defined by integrals. When a continuous function f that does not possess an elementary function antiderivative arises in applications and yet its antiderivative is needed, the Fundamental Theorem tells us that the "special function" antiderivative, $F(z)$, is well-defined by

$$F(z) = c \int_a^z f(x)\, dx. \tag{19}$$

Here, the lower limit a and constant c are chosen as some fixed values, compatible with the particular historical tradition for the specific F.

Log.—We've seen that the "cleanest" way to define the exponential and logarithmic functions is to begin with the definition

$$\ln z = \int_1^z \frac{dx}{x}. \tag{20}$$

Exercise 1 Identify f, F, a, and c in Equation 19 definition for the log function as defined in Equation 20.

Most calculus texts show how all the familiar properties of the log follow from the integral definition. The next problem gives a start on this program.

Exercise 2 From Equation 20 evaluate $\frac{d}{dz}\ln(z)$ and $\ln 1$.

Remark: *Maple*'s built-in natural ln function is written `ln(x)` or `log(x)` and the log base b ($\log_b x$) is written `log[b](x)`. For the special base of 10, one can also write `log10(x)`.

Erf.—The special function, erf, arises in connection with the probability

$$P(z) = \frac{1}{\sqrt{2\pi}} \int_{-z}^z e^{-x^2/2}\, dx = \sqrt{\frac{2}{\pi}} \int_0^z e^{-x^2/2}\, dx \tag{21}$$

of being no more than z standard deviations from the mean in a normal distribution:

```
P := sqrt(1/Pi) * int( exp(-x^2/2), x = 0..z);
                        1/2              1/2
              P := 1/2 2     erf(1/2 2     z)
```

The function erf is defined as a multiple of the antiderivative of e^{-x^2}. Erf is not an elementary function and is defined by:

$$\mathrm{erf}(z) = \frac{2}{\sqrt{\pi}} \int_0^z e^{-t^2}\, dt. \tag{22}$$

Exercise 3 What are f, F, a, and c in Equation 19 for the erf function as defined in Equation 22?

Exercise 4 Use Equation 22 and a change of variables to prove that *Maple*'s output above for P (as defined by Equation 21) is correct.

Exercise 5 Evaluate erf(0) and $\frac{d}{dz}$erf(z).

    The Bible for Special Functions.—Milton Abramowitz and Irene A. Stegun edited the *Handbook of Mathematical Functions with Formulas, Graphs and Mathematical Tables* under the auspices of the National Science Foundation and Massachusetts Institute of Technology. This work is cited a *vast* number of times in the scientific literature and you may want to consider buying your own copy of the Dover Edition.

    The Handbook begins with four chapters on the elementary matters and then in succeeding chapters discusses the "special functions" common in scientific applications. In case you are skeptical about the importance of defining functions by integrals, let's look at some of the items in the early chapters of the Handbook:

Chapter 5: Exponential Integral and Related Functions

Exponential Integral $\mathrm{E}_1(z) = \int_z^\infty \frac{e^{-t}}{t}\, dt$

Sin Integral $\mathrm{Si}(z) = \int_0^z \frac{\sin t}{t}\, dt$

Sinh Integral $\mathrm{Shi}(z) = \int_0^z \frac{\sinh t}{t}\, dt$

Chapter 6: Gamma Function and Related Functions

Gamma Function $\Gamma(z) = \int_0^\infty t^{z-1} e^{-t}\, dt$

Incomplete Beta Function $\mathrm{B}_x(a,b) = \int_0^x t^{a-1}(1-t)^{b-1}\, dt$

Chapter 7: Error Function and Fresnel Functions

Error Function $\mathrm{erf}(z) = \frac{2}{\sqrt{\pi}} \int_0^z e^{-t^2}\, dt$

Complementary Error Function $\mathrm{erfc}(z) = \frac{2}{\sqrt{\pi}} \int_z^\infty e^{-t^2}\, dt$

Cosine Integral $\mathrm{C}(z) = \int_0^z \cos(\frac{\pi}{2}t^2)\, dt$

Sine Integral $\mathrm{S}(z) = \int_0^z \sin(\frac{\pi}{2}t^2)\, dt$

And so it goes for 20-odd more chapters!

IN THE LAB

Exercise 6 Do the computations indicated in Exercise 2 using *Maple*'s built-in natural log function.

Exercise 7 Make a table (e.g. using the `for` statement) with three columns. In the first column, show the numbers between 1 and 2 at spacing 0.1, in the second column show the corresponding values of ln as defined by Equation 20, and in the third column show the values of *Maple*'s `ln`.

Exercise 8 Derive the sixth degree approximation, $P_6(z)$ for erf(z). Check that the first term agrees with the result you got in Exercise 5.

Exercise 9 Make a table showing values between 0 and 2 at spacing 0.25 in the first column, the corresponding values of erf as defined by Equation 22 in the second column, the values of *Maple*'s `erf` in the third column and the approximate values given by your result in Exercise 8 in the fourth and final column.

Exercise 10 Evaluate erf(∞) with both your direct integral definition (Equation 22) and with *Maple*'s built-in function. In *Maple*, ∞ is written `infinity`.

Exercise 11 Hand in a plot of erf on the interval $[-5, 5]$ and discuss how your plot bears out the other results you've obtained for the erf function, in particular, those of Exercises 5 and 10.

AFTER THE LAB

Exercise 12 Abramowitz and Stegun cite the following series for the erf function:

$$\text{erf}(z) = \frac{2}{\sqrt{\pi}} \sum_{n=0}^{\infty} \frac{(-1)^n z^{2n+1}}{n!(2n+1)}.$$

Recalling that conventionally $0! = 1$, write out the first few terms of the series explicitly and verify that they agree with the approximation you got in Exercise 8.

BEFORE THE LAB

Polar plotting is not in the pre-loaded packages, but you can make the `polarplot` routine available with the following command:

```
with(plots, polarplot);
```

One then plots the polar function $r = f(\theta) = \sin 2\theta$ like this:

```
polarplot( sin(2*theta), theta = 0..2*Pi );
```

The result is shown in Figure 26—it is called a *four leaved rose.*

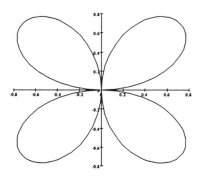

FIG. 26. `polarplot( sin(2*t), t = 0..2*Pi )`

Remark: Henceforth we shall use the shorthand t instead of `theta` in our plotting commands to save typing.

It is surprisingly difficult to predict the behavior of polar plots. For example, in the present example, since the period of $\sin 2\theta$ is π, we might think it sufficient to plot in the smaller range, 0 to π. But then we only get the right half of the graph, see Figure 27. And if $\sin 2\theta$ produces a "four leaved rose," then surely $\sin 3\theta$ produces 6 leaves? No, three leaves and this time it is all done at π instead of 2π, see Figure 28. We hope you enjoy using *Maple* to explore and understand the nature of polar plots related to the ones we've just shown you—such a project would be out of bounds if we couldn't produce the plots automatically. Moreover, you should know that the "roses" come up in antenna design, earthquake radiation patterns, etc. So they are useful, as well as beautiful.

IN THE LAB

Exercise 1 Experiment with `polarplot` on the polar equation $r = \sin k\theta$, $k = 1, 2, 3, \ldots$ and find out:

a) How the plot range needed to get a complete graph depends on the integer k.

b) How the number of "leaves" depends on the integer k.

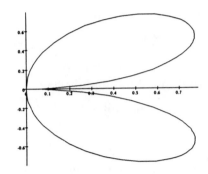

FIG. 27. polarplot(sin(2*t), t = 0..2*Pi)

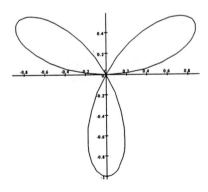

FIG. 28. polarplot(sin(2*t), t = 0..2*Pi)

Remark: Figures 26 and 28 provide correct information; your results should be consistent with them. Also, don't forget about the case $k = 1$!

Exercise 2 For $r = \sin\left(\dfrac{\theta}{k}\right)$, with k still a positive integer, find out how the plot range needed to get a complete graph depends on k.

Exercise 3 (**Harder**) For $r = \sin\left(\dfrac{m\theta}{n}\right)$, with m and n positive integers, find out how the plot range needed to get a complete graph depends on m and n. **Hints:** You can assume that m and n have no common divisors. You already have some evidence from Exercises 1 and 2. **Remark:** The full polar plot of something like $\sin\left(\dfrac{6\theta}{7}\right)$ is lovely, we hope you enjoy seeing it.

Exercise 4 Describe what happens if we use an irrational factor b in $r = \sin b\theta$. In particular, for $b = \sqrt{2}$ plot this function on each of the intervals $[0, 12\pi]$, $[0, 24\pi]$ and $[0, 48\pi]$.

Exercise 5 Investigate the polar plots of the family of functions given by $r(\theta) = 1 + p\sin\theta$ for various values of the *positive* parameter p. Begin by trying the values $p = 3/4$, 1, and $4/3$. Describe how the family looks in terms of the p value. In particular, tell what happens as $p \to 0$ and as $p \to +\infty$.

Exercise 6 Hand in the most interesting polar plot you can create. Think about using exponentials, powers of trig functions, etc. Just to get your juices going, try $r(\theta) = 3\cos^2\theta - 1$ and $3\cos^3\theta - 1$, but we're sure you can do a lot better!

AFTER THE LAB

Exercise 7 By consulting your text, put names on the plots you made in Exercise 5.

Exercise 8 From your results in Exercise 5, describe how the two parameter family $r(\theta) = a + b\sin\theta$ looks for various values of the *positive* parameters a and b.

BEFORE THE LAB

Analytical Geometry

Analytical Geometry uses the power of algebra to prove geometric theorems. Ideally, it reduces the proof to a straightforward calculation. Choosing the right coordinate system and location of the geometric quantities simplifies these calculations and reveals the true elegance of the method.

Illustrative Example Find the locus of all points equidistant from two given fixed points. **Remark**: The word "locus" means the set of points satisfying a given condition. One might describe a locus by the equation of the points satisfying the condition (algebra) or by giving a spatial description (geometry).

Answer: Choose the two given points (say P_1 and P_2) to be on the x-axis and choose their midpoint to be the origin. That is, pick the points as $(c, 0)$ and $(-c, 0)$, where c is a fixed parameter such that $2c$ is the distance between the points (see Figure 29). Denote the

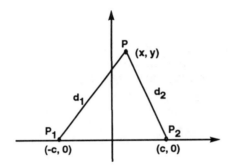

FIG. 29. Geometry for the Illustrative Example.

coordinates of an arbitrary point P on the desired locus by (x, y) and construct the distances d_1 and d_2 using the Pythagorean theorem as $d_1 = \sqrt{(x+c)^2 + y^2}$ and $d_2 = \sqrt{(x-c)^2 + y^2}$. The given condition is then expressed as $d_1 = d_2$, after squaring we obtain the equation

$$x^2 + 2cx + c^2 + y^2 = x^2 - 2cx + c^2 + y^2.$$

We see that all the squared terms cancel and so, after transposing, we obtain $4cx = 0$. Thus, the equation of the required locus is $x = 0$ (algebraic description) which can also be recognized as the perpendicular bisector of the seqment $\overline{P_1 P_2}$ connecting the given points (geometric description). Note: The (rather obvious) result shows that the point P in Figure 29 should have been put on the y-axis, but in drawing the picture, it wouldn't have been appropriate to assume the result ahead of time. Finally, in anticipation of more difficult problems, let's use *Maple* to help with the algebra:

```
d1 := sqrt( (x+c)^2 + y^2 ):
d2 := sqrt( (x-c)^2 + y^2 ):
expand( d1^2 - d2^2 );
                    4 x c
```

So again, we see that $d_1^2 = d_2^2$ implies that $4cx = 0$.

The collect command

The collect command can be used to group messy expressions according to the variables we specify as significant:

```
q := c^2 - k^2 * c^2 + x^2 -a^2 * x^2 + y^2 - b^2 * y^2:
collect( q, [x,y,c] );
                            2    2            2    2           2   2
              (1 - a ) x  + (1 - b ) y  + (1 - k ) c
```

Plotting Implicit Functions in *Maple*

An implicit function of two variables is one in the form $F(x, y) = 0$. To graph such a function at a given value $x = x_0$, we conceptually have to solve for y. Since this might be hard to do, plotting implicit functions is also hard. However, in the auxiliary packages, *Maple* provides a function that tries to accomplish the job. Here is an example:

```
with(plots):
implicitplot( ((x-1)/3)^2 + y^2 = 8, x = -10..10, y = -10..10,
    scaling = CONSTRAINED);
```

First, using with, we call in the auxiliary graphics packages (do only once in a session) and then we make the plot of the ellipse shown in Figure 30. Note: the scaling = CONSTRAINED option prevents Maple from rescaling and, in this case, essentially drawing a circle.

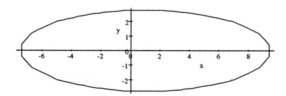

FIG. 30. Result of implicitplot.

IN THE LAB

Exercise 1 Extend the Illustrative Example to the case where the points on the locus are k times as far from one fixed point as they are to the other. Method: Use the same set up as above, express the locus as $d_1 = k \cdot d_2$ and show that the problem reduces to simplifying the expression $d_1^2 - k^2 d_2^2$. Use *Maple* to get the locus into recognizable form (don't forget about `collect` and don't be hesitant to do a little extra post-simplification by hand). Record the equation for the locus for later use.

Exercise 2 Use `implicitplot` to find out how the locus you obtained in Exercise 1 varies with k (take $c = 1$). In particular, explore the behavior as $k \to 0$, 1, and $+\infty$. Print out, or carefully sketch, the key graphs.

Exercise 3 Show that the locus of all points whose *sum* of distances from two fixed points is constant is an ellipse (if necessary, consult the index of your text for the equation of an ellipse). Method: Use the same set up as above, expressing the locus as $d_1 + d_2 = $ constant $= 2a$. (The 2 is traditional and makes the final equation nicer.) Now square as before:

$$d_1^2 + 2d_1 d_2 + d_2^2 = 4a^2.$$

And now we see a problem—the term $2d_1 d_2$ still involves square roots. So we have to isolate this term by transposing and squaring again. This gives

$$4d_1^2 d_2^2 = \left(4a^2 - d_1^2 + d_2^2\right)^2.$$

Transpose and use *Maple* to simplify the resulting formidable expression and thus obtain the requested result. **Remark**: If you like algebra, you will enjoy doing the simplification by hand to see how all the miraculous cancellations work out.

Exercise 4 Show that the locus of all points whose *difference* of distances from two fixed points is constant is a hyperbola. Think before you compute.

Exercise 5 We can construe the locus of Exercise 1 as specifying that the *quotient* of distances from two fixed points be constant (i.e., $d_1/d_2 = k$). Thus, with Exercises 3 and 4, we've treated sums, difference and quotients from two fixed points being constant. So naturally, we want you to now do the missing case! Investigate the same problem for *products* (i.e., $d_1 d_2 = k$). Is the result a conic like the other three cases? (A conic may be defined as a locus satisfying an equation in x and y of second degree.) Use `implicitplot` to graph the fixed product locus (take $c = 1$ and vary k). Make sketches or hand in key graphs that illustrate your mastery of this locus's behavior. Suggestion: You will probably get nicer pictures if you change all the 10's to 4's in the `implicitplot` command we've been using.

AFTER THE LAB

Exercise 6 By completing the square in the locus you found in Exercise 1, explain the behavior you discovered in Exercise 2.

BEFORE THE LAB

The equations $x = \cos t$, $y = \sin t$ with the parameter t in the range $[-\pi, \pi]$, gives a familiar parametric representation of the unit circle (note that $x^2 + y^2 = 1$). Here is a *Maple* code to get a plot of a parametric form:

```
x := t -> cos(t):
y := t -> sin(t):
plot( [x, y, 0..2*Pi] );                        # or:  plot( [x(t), y(t), t = 0..2*Pi] );
```

The result is shown in Figure 31.

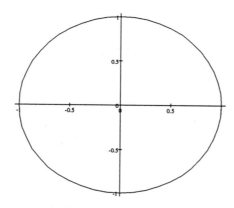

FIG. 31. A parametric plot of the unit circle.

IN THE LAB

The unit circle

The parameterization of the unit circle given above leads to a transversal of the circle with unit speed:

$$
\begin{aligned}
v(t) &= \sqrt{(x'(t))^2 + (y'(t))^2} \\
&= \sqrt{(-\sin t)^2 + (\cos t)^2} \\
&= 1.
\end{aligned}
$$

Or using *Maple*:

```
v := t -> simplify( sqrt( D(x)(t)^2 + D(y)(t)^2 ) );
                        1
```

Exercise 1 Show that the parametric equations $x(t) = (1 - t^2)/(1 + t^2)$, $y(t) = 2t/(1 + t^2)$ give an alternate representation of the unit circle (technically, the point $(-1, 0)$ is "missing"). Make a table showing the values of x, y and the speed v for the points $t = -5, -4, \ldots, 3$, 4, and 5. Describe in words how the speed varies as the point moves around the unit circle. Discuss the "missing point" in terms of limits.

The Hypocycloid

The *hypocycloid* is the curve generated by a circle of radius b rolling on the inside of a larger circle of radius a, see Figure 32. The equations of the hypocycloid are given

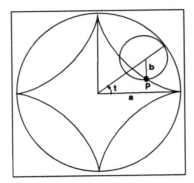

FIG. 32. A hypocycloid with $b/a = 1/4$.

parametrically as:

$$x(t) = (a - b)\cos t + b\cos\frac{a - b}{b}t,$$

$$y(t) = (a - b)\sin t - b\sin\frac{a - b}{b}t, \qquad a > b.$$

We could factor out the scaling factor a in each equation and study the family of hypocycloids in terms of the single parameter b/a. Alternately, we just set $a = 1$ in the following *Maple* code, but label our plots in terms of the ratio:

```
x := t -> (a-b)*cos(t) + b*cos( (a-b)*t/b ):
y := t -> (a-b)*sin(t) - b*sin( (a-b)*t/b ):
a := 1:  b := 1/3:
plot( [x, y, 0..2*Pi] );
```

The output of these commands is shown in Figure 33.

Exercise 2 Do graphical experiments with the above code to determine the period of the hypocycloid for the case when $a = 1$ and $b = 1/k$, with $k = 3, 4, 5, \ldots$. Write a sentence describing the nature of the graph.

Exercise 3 Continuing the previous problem, describe what happens for $a = 1$, $b = 1/k$, as $k \to \infty$.

Exercise 4 Determine the arclength of the limiting curve in the previous problem. **Hint:** most approaches go better if you get the arclength for one "lobe" and multiply by the number of lobes. (You may have to help *Maple* out with some obvious algebra before integrating).

Exercise 5 Do graphical experiments with the given hypocycloid code to determine its period when $a = 1$ and $b = j/16$, for $j = 1, 3, 5, \ldots, 15$. Write a paragraph describing what you observe about this sequence of plots.

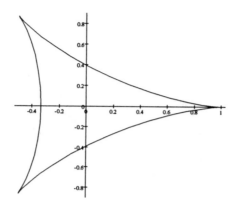

FIG. 33. A hypocycloid with $b/a = 1/3$.

Exercise 6 Perform whatever experiments you find necessary to determine the period of the hypocycloid for $a = 1$ and $b = j/k$ where j and k are integers with $j < k$.

Exercise 7 Hypocycloid pairs corresponding to j/k's given by $1/3$ and $2/3$, or $1/4$ and $3/4$, or in general $1/k$ and $1 - 1/k$ have the same graph, but there is a difference, what is it? Be as quantitative as possible.

Exercise 8 Investigate the special hypocycloid corresponding to $a = 1$ and $b = 1/2$.

Another family of parametric curves

Exercise 9 Study the family $x(t) = \cos^k t$, $y(t) = \sin^k t$, for $k = 1, 2, \ldots$. Describe the nature of the graphs and estimate the arclength of the limiting curve as $k \to \infty$. Can you explain your result intuitively? (Again, you may have to help *Maple* with the simplifications before integrating).

AFTER THE LAB

Exercise 10 Look up (or derive yourself) the equations given above for the hypocycloid and give a clear and detailed exposition.

Exercise 11 Verify the conclusion you drew in Exercise 6 directly from the parametric equations of the hypocycloid.

BEFORE THE LAB

The standard parametric representation of the interior of the sphere with radius a is:

$$
\begin{aligned}
x &= r\cos\theta\sin\phi, & & & 0 \le r \le a, \\
y &= r\sin\theta\sin\phi, & \text{with} & & 0 \le \phi \le \pi, \\
z &= r\cos\phi, & & & 0 \le \theta \le 2\pi.
\end{aligned}
\tag{23}
$$

Exercise 1 Verify that Equations 23 satisfies the equation of the solid sphere,

$$
x^2 + y^2 + z^2 = r^2, \qquad 0 \le r \le a.
$$

To obtain the space curve from Equations 23, we first make r constant, say $r = 1$, to obtain the parametric representation of the *surface* of the unit sphere. Then to obtain a parametric representation of a *curve* on the surface of the unit sphere, we specify ϕ and θ in terms of a single, common parameter t. Let us choose a curve that starts at the North pole of the sphere, winds around the sphere n times and then finishes at the South Pole. To do this, we set $\phi = t$ and give t the range $[0, \pi]$. This gets us from the North Pole to the South Pole. To get the n twists around the sphere on the way down, we make θ proportional to t—in fact, we take $\theta = 2nt$. The 2 is needed because we have confined t to the $[0, \pi]$ range, whereas θ needs a 2π range to twist around once. Thus, we have the following parametric representation of our twisting sphere curves:

$$
\begin{aligned}
x &= \cos 2nt \sin t, \\
y &= \sin 2nt \sin t, \\
z &= \cos t,
\end{aligned}
\tag{24}
$$

with $0 \le t \le \pi$ and $n = 0, 1, 2, \ldots$, governing the number of twists (or orbits) in our space curve.

Exercise 2 Verify that for any value of n, the curve specified by Equations 24 lies on the unit sphere.

Exercise 3 Verify that the parametric equations

$$
\begin{aligned}
x &= \cos 2nt \sin t, \\
y &= \sin 2nt \sin t, \\
z &= 3 \cos t,
\end{aligned}
\tag{25}
$$

with $0 \le t \le \pi$, correspond to curves twisting down the ellipsoid

$$
x^2 + y^2 + \left(\frac{z}{3}\right)^2 = 1.
$$

Graphing space curves with *Maple*

To graph space curves, we call on **plots[spacecurve]**. Here is an example that plots Equations 24 for the case of three twists:

```
x := cos(2*n*t)*sin(t):
y := sin(2*n*t)*sin(t):
z := cos(t):
n := 3:
plots[spacecurve]( [x, y, z], t = 0..Pi, axes = BOXED,
     orientation=[45, 60] );
n := 'n':     # It's a good idea to clean up temporary variables
```

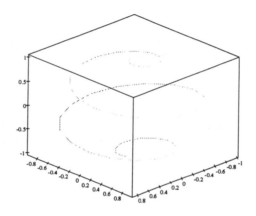

FIG. 34. Three twists down the sphere.

The output is shown in Figure 34. Notice that we have used the **orientation** option. It may be true that the single most important issue in plotting a three dimensional surface or curve is specifying where you want to look at the surface from. In the above example, we set $\theta = 45$ degrees and $\phi = 60$ degrees. Usually it takes quite a bit of experimenting to get the most satisfying viewing angle. You may find Table 1 helpful.

Table 1. Typical choices for the **orientation** option. Location of the 'eye'.

0, 90	directly in front
45, 45	(default) right and up
0, 45	in front and up
-45, 45	left and up

View points slightly above the surface usually work well. Having a box explicitly drawn around the curve or surface is helpful in recognizing the object, hence the axes = BOXED option.

IN THE LAB

Exercise 4 Change the `orientation` setting in the plotting command for our twisting sphere curves and hand in one or more plots that you think capture its behavior best. Repeat for $n = 4$.

Arclength of space curves

We now try our hand at computing the arclengths of the twisting sphere curves. Unfortunately, as is typical for arclength integrals, the exact answer cannot be obtained:

```
ds := simplify( sqrt( diff(x,t)^2 + diff(y,t)^2 + diff(z,t)^2 ) );
                        2       2        2      1/2
            ds := (4 n  - 4 n  cos(t)  + 1)
```

```
int( ds, t = 0..Pi );
        Pi
        /
        |      2       2        2      1/2
        |   (4 n  - 4 n  cos(t)  + 1)      dt
        |
        /
        0
```

We see that *Maple* returns an unevaluated integral, suggesting that it is unlikely that an elementary antiderivative exists. Thus we turn to *numerical* evaluations:

```
for n to 6 do
   evalf( Int( ds, t = 0..Pi ) )
od;
                        5.270367163
                        8.814275563
                        12.61161680
                        16.49505747
                        20.41852184
                        24.36404151
```

Exercise 5 It looks like the arclengths of the twisting sphere curves are approaching $4n$ where n is the number of twists. Investigate this for $n = 2^k, k = 1, 2, \ldots, 8$. Does the approximation improve with increasing n or does it "fall apart?"

Exercise 6 Repeat Exercise 5 for the curves defined by Equations 25.

Exercise 7 Let $\mathcal{C}$ be a cylinder of unit radius whose base is the circle $x^2 + y^2 = 1$ in the xy-plane and whose top is at height 2π. We claim that a parametric representation of a curve that starts at the point $(1, 0, 0)$ winds n times around the cylinder and ends right above the starting point at $(1, 0, 2\pi)$ on $\mathcal{C}$ is given by:

$$x = \cos nt,$$
$$y = \sin nt,$$
$$z = t.$$

Use `plots[spacecurve]` to make a few graphs that verify the claim.

AFTER THE LAB

Exercise 8 The formula obtained for the differential arclength in Exercise 5 was

$$ds = \sqrt{1 + 2n^2(1 - \cos 2t)}\, dt.$$

Use the fact that 1 is negligible compared to n^2 for n large to explain the numerical results of that Exercise. **Hint**: After dropping the "1," exercise those good old trigonometric identities to write

$$1 - \cos 2t = 2\sin^2 t.$$

Exercise 9 Consider the case $n = 0$ for the twisting sphere curves given by Equations 24. Describe the resulting curve and compute its arclength. Is the result compatible with the result of Exercise 5? Explain.

Exercise 10 Repeat Exercise 8 for the curves defined by Equations 25 (also see Exercise 6).

Exercise 11 Give a convincing argument for the correctness of the claim made in Exercise 7.

Exercise 12 Compute the arclengths of the parametric curves of Exercise 7. Verify that your answer makes sense for $n = 0$. Give an approximation valid for large n. Can you give a geometric interpretation of the meaning of this approximation?

Vectors in Mathematica

Here is the *Maple* notation for vectors:

```
with(linalg):          # only necessary once
v := [2, 3, 4]:
w := [1, 1, 1]:
```

Operations with them require some care:

```
add(v, w);
                              [ 3, 4, 5 ]

u := add(9*v, -29*w);
                          u := [ -11, -2, 7 ]

dotprod(v, w);
                                   9

dotprod(u, v);          # note these two vectors are perpendicular
                                   0
```

Test your understanding of what the above statements do by checking them!

It is also possible to define vector *functions*. For example, the vector function **r** of the scalar variable t representing the helical space curve,

$$\boldsymbol{r}(t) = \cos t\,\boldsymbol{i} + \sin t\,\boldsymbol{j} + t\,\boldsymbol{k}, \qquad (26)$$

is rendered in *Maple* as

```
r := [cos(t), sin(t), t] ;
```

We can then compute the corresponding velocity vector like this:

```
v := diff(r, t);
                    v := [- sin(t), cos(t), 1]
```

Recall the famous inverse square law, expressing the attractive force between masses at the origin and **r**:

$$\boldsymbol{F}(\boldsymbol{r}) = k\frac{\boldsymbol{r}}{r^3}. \qquad (27)$$

Here, **r** represents the position vector,

$$\boldsymbol{r} = x\,\boldsymbol{i} + y\,\boldsymbol{j} + z\,\boldsymbol{k},$$

r represents the length of this vector. The inverse square law is rendered in *Maple* as

```
F := k*(x,y,z)/(x^2 + y^2 + z^2)^(3/2);
```

As an example, here is one way to evaluate the inverse square law force on a particle moving on the helix of Equation 26:

```
F := k * [ r[1], r[2], r[3] ] / ( r[1]^2 + r[2]^2 + r[3]^2 )^(3/2);

                        [cos(t), sin(t), t ]
        F := k  -------------------------------------
                         2        2      2 3/2
                   (cos(t)  + sin(t)  + t )
```

IN THE LAB

Exercise 1 Use *Maple*'s dot product to help find a value of the parameter a such that the following three dimensional vectors, u and v are "orthogonal" (i.e., perpendicular):

$$u = [1,2,3], \quad \text{and} \quad v = [5,8,a].$$

Same question for the vectors:

$$u = [1,2,3,4], \quad \text{and} \quad v = [2,5,8,a].$$

Remark: The last pair of vectors were 4-dimensional. Indeed, 4-D, 5-D, ..., vectors *do* occur in practical and very important applications. We can operate with them by analogy, even though we can't fully visualize fourth and higher dimensional objects).

The Euclidean *norm* or length of a vector u is defined as $\sqrt{u \cdot u}$. However, *Maple* uses a different norm which has its advantages, namely: the absolute value of the largest component. For example, let $u = [2,-3,1]$. Then the Euclidean norm, call it *norm2* is: $\sqrt{4+9+1} = \sqrt{14}$.

But *Maple*'s norm, accessed by the command `norm(u)`, is clearly 3.

Exercise 2 Use *Maple* to find the length of the 2-D vector, $[1,1]$. Repeat for $[1,1,1]$, $[1,1,1,1]$ and $[1,1,1,1,1]$. (Again, proceed by analogy for the 4-D and 5-D vectors without worrying about the geometrical interpretation.)

A vector can be scaled into a *unit* vector by dividing by its Euclidean norm; using norm2:

```
u := [1, 2, 3]:
norm2 := sqrt( dotprod(u,u) ):
uhat := u / norm2;
                                              1/2
                      uhat := 1/14 [1, 2, 3] 14
dotprod(uhat,uhat);      # check it:
                              1
```

Exercise 3 Use *Maple* to find the unit vectors that have the same directions as the three vectors $[1,1,1]$, $[1,-2,1]$, and $[3,0,-3]$. Show that the resulting unit vectors are mutually orthogonal. Show that the original vectors were also mutually orthogonal. Does one of these orthogonality results *always* imply the other?

Exercise 4 For the vector giving the position vector on the helix as defined in Equation 26, use *Maple* to compute the corresponding velocity vector v and acceleration vector a. Which of r, v, and a are orthogonal?

The work integral

As you know, in one dimension, work is defined as

$$W = \int_a^b F(x)\,dx,$$

where F denotes the force applied to a particle moving on the x-axis from $x = a$ to $x = b$. Formally, the generalization to three (or any number of) dimensions is

$$W = \int_C \boldsymbol{F}(\boldsymbol{r}) \cdot d\boldsymbol{r},$$

where the particle now moves on the curve C under the influence of the force $\boldsymbol{F}$. However, this abstract definition takes on concrete meaning only when we introduce a definite representation for C, for example, the parametric representation,

$$\boldsymbol{r}(t) = x(t)\,\boldsymbol{i} + y(t)\,\boldsymbol{j} + z(t)\,\boldsymbol{k}.$$

Then the work integral can be written as an ordinary one dimensional integral over t and all mystery evaporates:

$$W = \int_{t_0}^{t_1} \boldsymbol{F}(\boldsymbol{r}(t)) \cdot \boldsymbol{r}'(t)\,dt. \tag{28}$$

Here, t_0 and t_1 respectively denote the values of t corresponding to the beginning and end of the curve C. A typical *Maple* rendition of this integral of a dot product is

```
rprime := diff(r, t):
int( dotprod(F, rprime), t = 0..2*Pi );
    # or use evalf( Int(...))    for numerical integration
```

Exercise 5

a) Compute the work as given by Equation 28 on the helical path given by Equation 26 when the force is the near surface gravity vector, $\boldsymbol{F} = mg\,\boldsymbol{k}$. Take the mass moved as $m = 1000\,\text{kg}$ and take the surface gravitational acceleration as $g = 0.0098\,\text{km/sec}^2$.

b) Repeat the work computation using the inverse square law given by Equation 27. To get the helix located on the earth's surface, assume that the helix is located at the North pole and thus add the earth's radius R to the z-component of the helix . At the earth's surface, the near surface approximation and the inverse square law must agree. Thus the constant k is determined by $k/R^2 = mg$ or $k = mgR^2$. Use the value $R = 6400\,\text{km}$ for the earth's radius.

Exercise 6 Exercise 5 has the flavor of computing the work done in ascending a hill. However, the helical path is not very realistic. Here's our attempt at a more plausible path:

$$
\begin{aligned}
x &= e^{-t/3}\cos 3t, \\
y &= e^{-t/3}\sin 3t, \\
z &= 13t/(t^2 + 40),
\end{aligned}
$$

with $0 \le t \le 2\pi$.

a) Do a `plots[spacecurve]` with a well chosen `orientation` and comment on whether the given curve *does* represent a more realistic hill climb than the helix. Review the `plots[spacecurve]` command in the *Twisting Space Curves* project if necessary.

b) Compute the work for the near surface gravity approximation.

c) Compute the work for the inverse square law.

AFTER THE LAB

Exercise 7 Discuss the numerical results in Exercises 5 and 6.

BEFORE THE LAB

In this project, we return to our roots[2]. In our first serious project, *Graphical Equation Solving*, we solved cubic equations (i.e., found their roots) by plotting them and "zooming in" on the intersections with the x-axis. The central problem of that project was determining the normals to the parabola $y = x^2$ from a given point (p, q). We are going to generalize that problem to three-dimensions, so please re-read the Illustrative Example and Exercises 4, 5, and 6 of the *Graphical Equation Solving* project. Exercises 7, 10, and 11 of that project also are about the normal problem, but are of secondary importance to us now.

Your knowledge has increased greatly since you did that early lab project: you have mastered the derivative, the integral, Taylor Series and much more—you have a lot to be proud of! In particular, you now know all about Newton's Method for solving equations. In this project we will generalize Newton's Method from one equation in one unknown to two equations in two unknowns. So also please re-read the *Newton's Method* project, paying particular attention to the `for` loop implementation of this algorithm given there. Some texts give a careful derivation of Newton's Method in two dimensions—if yours does, you can just skim read the sketch of the derivation and interpretation given next and go on to the *Illustrative Exercise*.

Recall the derivation of Newton's Method in *one* dimension. Given a function $y = f(x)$ and a starting value x_0 near a root (or zero) of the function (i.e., $f(x_0) \approx 0$), we seek a better approximation to the root. The idea of Newton's method is to replace the function by its linear approximation at the starting value, thus obtaining $y \approx f(x_0) + f'(x_0)(x - x_0)$. Then we put $y = 0$ and solve this linear equation for x to get $x \approx x_0 - f(x_0)/f'(x_0)$. We take this approximate solution as our next starting point

$$x_1 = x_0 - f(x_0)/f'(x_0)$$

and repeat the process to obtain the general Newton iteration

$$x_{n+1} = x_n - f(x_n)/f'(x_n).$$

We know that if there really is a solution near (p, q), Newton's method usually converges rapidly to that solution. (We also know that there are exceptional cases where Newton's method diverges!)

An alternate road to the Newton iteration result is to write the equation of the tangent line at $(x_0, f(x_0))$ as

$$y - f(x_0) = f'(x_0)(x - x_0)$$

and then solve for the point where the tangent line intersects the x-axis by putting $y = 0$ and denoting the intersection point by $x = x_1$. In any case, recall that in the *Maple* `for` loop implementation, we overwrite the old guess with the new one at each iteration:

---

[2]Most puns are awful and this one is no exception to the rule.

```
f  :=  x ->  # your function goes here
xn :=        # your starting guess goes here
for n to k do
   xn := xn - f(xn) / D(f)(xn)
od;
```

Now consider the problem in *two* dimensions. Suppose we have the two equations

$$z = f(x, y),$$
$$z = g(x, y).$$

and a starting value (p, q) near a simultaneous root of the equations (that is, both $f(p, q) \approx 0$ and $g(p, q) \approx 0$). Again, we seek a better value by using the linear approximation to the equations:

$$z \approx f(x_0, y_0) + f_x(x_0, y_0)(x - x_0) + f_y(x_0, y_0)(y - y_0)$$

$$(29)$$

$$z \approx g(x_0, y_0) + g_x(x_0, y_0)(x - x_0) + g_y(x_0, y_0)(y - y_0).$$

Put $z = 0$ in each of these approximate equations and then solve for (x, y). If we denote the solution by (x_1, y_1), then we obtain the (usually) improved approximation:

$$x_1 = x_0 - \frac{f(x_0, y_0)g_y(x_0, y_0) - g(x_0, y_0)f_y(x_0, y_0)}{f_x(x_0, y_0)g_y(x_0, y_0) - g_x(x_0, y_0)f_y(x_0, y_0)}$$

$$(30)$$

$$y_1 = y_0 - \frac{g(x_0, y_0)f_x(x_0, y_0) - f(x_0, y_0)g_y(x_0, y_0)}{f_x(x_0, y_0)g_y(x_0, y_0) - g_x(x_0, y_0)f_y(x_0, y_0)}$$

We'll give the *Maple* code for this 2D iteration scheme below.

Illustrative Exercise: *Find the equation of the line through the point $P(4, 1, 0)$ that is normal to the paraboloid $z = x^2 + 4y^2$.*

Solution: We know that it is good scientific practice to solve the most *general* problems we can, so rather than doing algebra that has to be done all over again when the surface or the point is changed, we immediately generalize the stated problem to:

Find the equation of the line through the point $P(p, q, r)$ that is normal to the surface $z = h(x, y)$.

Introduce the point where the normal line intersects the surface as $(a, b, c) = (a, b, h(a, b))$. The parametric form of a line in 3-space is: $r = r_0 + nt$. Here, r_0 denotes the given point, so that $r_0 = [p, q, r]$. Furthermore, n denotes the normal to the surface at the point $(a, b, h(a, b))$, so that $n = [h_x(a, b), h_y(a, b), -1]$. Finally, r represents a vector to the general point on the normal. So, in particular, we can take it to be the point on the surface: $r = [a, b, h(a, b)]$. With these replacements, the normal line equations give:

$$a = p + h_x(a, b)t$$
$$b = q + h_y(a, b)t$$
$$h(a, b) = r - t$$

We use the last equation to eliminate t, arriving at two equations for the two unknowns, a and b:

$$a = p + h_x(a, b)(r - h(a, b))$$

$$b = q + h_y(a, b)(r - h(a, b))$$

(31)

For the case of our paraboloid,

$$z = h(x, y) = x^2 + 4y^2,$$

Equations 31 reduce to (cf. Exercise 9):

$$2a^3 + 8ab^2 + (1 - 2r)a = p$$

(32)

$$32b^3 + 8ba^2 + (1 - 8r)b = q.$$

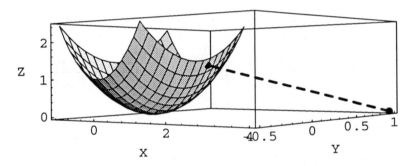

FIG. 35. Surface and normal for the Illustrative Exercise.

Now that we have the two nonlinear equations to solve for a and b, and have an iteration scheme in hand, we come up against the really hard question: *how do we find a good starting point?* In the case of a single equation, we could often make a plot of $f(x)$ that revealed a good approximation of the root(s), but in two dimensions, directly plotting the surface $h(x, y)$ in a way that reveals a good starting value is usually too time consuming to be considered. Figure 35 does a good job for our special case, but the truth is, we need to know the answer in order to pick a good `orientation` and good plot limits in x and y. Getting a good three dimensional picture by doing an honest search is out of the question for most surfaces and points.

Here is a *conceptual* approach—we are *not* suggesting that you actually carry it out: In place of three dimensional graphics, after filling in specific values for p, q, and r, we could try to make a sketch of each of the two dimensional curves in Equations 32. To do this, we could pick a b in one of the equations, and solve it for the corresponding a-value(s). Then pick another b and solve for a. Once we have enough points, we could plot that equation. Then we plot the other one the same way. The *intersections* of these two plots would give our starting point. Certainly a lot of work, but conceptually plausible because we could make use of our one dimensional plotting/Newton methodology in obtaining each a-value.

Since we understand how to proceed "in principle," in order to make this project practical to do, we will use `implicitplot` to get the two curves plotted and then pick starting values from the intersections we see. (This is usually preceeded by rough 3-Dim plots to get an idea of the region of interest; unless e.g. the physics of the problem gives this information). Figure 36 shows the results of the `implicitplot` command :

```
p := 4:  q := 1:  r := 0:
f := (a,b) -> 2*a^3 + 8*a*b^2 + (1 - 2*r)*a - p:
g := (a,b) -> 32*b^3 + 8*b*a^2 + (1 - 8*r)*b - q:
with(plots):      # only necessary once
implicitplot( {f(a,b) = 0, g(a,b) = 0}, a = 0..2, b = -1..1,
              scaling = CONSTRAINED );
```

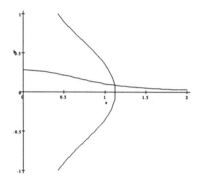

FIG. 36. `implicitplot` for the Illustrative Exercise.

Thus we get the starting value $a = 1.1$, $b = 0.1$. Here is the code to carry out the Newton iteration and the results of running that code. With the functions and parameters as above:

```
dfx := (a,b) -> D[1](f)(a,b):
dfy := (a,b) -> D[2](f)(a,b):
dgx := (a,b) -> D[1](g)(a,b):
dgy := (a,b) -> D[2](g)(a,b):
an := 1.1:  bn := .1:
for n to 8 do
   an := an - ( f(an,bn)*dgy(an,bn) - g(an,bn)*dfy(an,bn) ) /
             ( dfx(an,bn)*dgy(an,bn) - dgx(an,bn)*dfy(an,bn) ):
   bn := bn - ( g(an,bn)*dfx(an,bn) - f(an,bn)*dgx(an,bn) ) /
             ( dfx(an,bn)*dgy(an,bn) - dgx(an,bn)*dfy(an,bn) ):
   lprint( an, bn )
od;
1.120451160    8869571155.0E-11
1.119976895    8860526798.0E-11
1.119976715    8860526310.0E-11
1.119976715    8860526306.0E-11
1.119976715    8860526311.0E-11
```

```
1.119976715    8860526307.0E-11
1.119976715    8860526312.0E-11
1.119976715    8860526308.0E-11
```

Finally, we state our answer to the problem:
The normal line, $r = r_0 + nt$, is given by

$$
\begin{aligned}
x &= p + h_x(a, b)t = p + 2at \\
y &= q + h_y(a, b)t = q + 8bt \\
z &= r - t
\end{aligned}
$$

where the given point is $[p, q, r] = [4, 1, 0]$, and $a \approx 1.119976715$, $b \approx 0.08860526308$.

Exercise 1 Consider the problem of finding the point $x = a$, $y = b$ on the surface $z = h(x, y)$ that minimizes the distance between the surface and the given point $P(p, q, r)$. Show that a and b also satisfy Equations 31. State the meaning of this result.

IN THE LAB

Exercise 2 Find the normal line to the paraboloid in the Illustrative Exercise, when the given point is $(2, 2, 0)$.

Exercise 3 Find the minimum distance from the point $(2, 2, 0)$ of the previous Exercise to the paraboloid in the Illustrative Exercise. **Hint:** Given the result of Exercise 1, all you have to do is set up the formula for 3D distance and apply it appropriately.

Exercise 4 Find the normal line to the paraboloid in the Illustrative Exercise, when the given point is $(4, 1, 4)$. Warning: There is more than one root in this case, find all the roots (and hence all the normals) that you can.

Exercise 5 Which of the solutions in the last Exercise correspond to the *minimum* distance? What is the nature of the other solutions (i.e., are they local minima or something else?)

AFTER THE LAB

Exercise 6 Derive the one dimensional Newton iteration for x_1 by using the tangent line approach.

Exercise 7 Justify Equations 30 by solving the linear system

$$
\begin{aligned}
Ax + By &= r \\
Cx + Dy &= s
\end{aligned}
$$

and identifying the constants A, B, C, D, r, and s in Equations 29 (after putting $z = 0$ in those equations).

Exercise 8 Explain why Equations 30 can alternately be derived by finding the point (x_1, y_1), where the tangent planes

$$
\begin{aligned}
z - f(p, q) &= f_x(p, q)(x - p) + f_y(p, q)(y - q) \\
z - g(p, q) &= g_x(p, q)(x - p) + g_y(p, q)(y - q)
\end{aligned}
$$

to the functions f and g at (p, q) intersect the $z = 0$ plane.

Exercise 9 Make the necessary identifications to specialize Equations 31 for the case of the paraboloid,

$$
z = h(x, y) = x^2 + 4y^2,
$$

and show that the non-linear equations to be solved in this case reduce to Equations 32.

BEFORE THE LAB

In this project we'll show you how to do multiple integrals with *Maple*. As an application, we will compute the volumes and center of mass locations of some simple bodies in 2, 3, and 4 (!) dimensions.

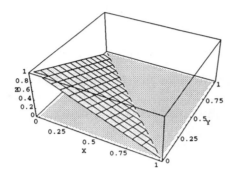

FIG. 37. Volume cut off by the plane $x + y + z = 1$ and the coordinate planes.

Illustrative Example 1. Find the volume of the region below the plane $x + y + z = a$ and satisfying: $0 \le x \le a$, $0 \le y \le a$, and $z \ge 0$.

Solution: The base of this region is the triangle in the xy-plane that is bounded by the coordinate axes and the line, $x + y = a$ (see Figure 37 in which we took $a = 1$). The mathematical expression for this volume is

$$\int_0^a \int_0^{a-x} \int_0^{a-x-y} 1 \, dz \, dy \, dx$$

and the rendition in *Maple* is:

```
volume := int( int( int( 1, z = 0..a-x-y ), y = 0..a-x ), x = 0..a );
                                  3
                  volume := 1/6 a
```

Exercise 1 Check the volume obtained above by a hand calculation.

Illustrative Example 2. Find the center of mass of the region in *Illustrative Example 1*.

Solution: To get the x component:

```
cmassX := int( int( int(x, z = 0..a-x-y ), y = 0..a-x ), x = 0..a ) / volume;
                      cmassX := 1/4 a
```

Similarly, cmassY = cmassZ = $a/4$.

Exercise 2 Check the center of mass obtained above by a hand calculation. Explain why the components of the center of mass are equal.

IN THE LAB

Exercise 3 Use *Maple* to compute the volume and center of mass of the region *above* the plane $x + y + z = a$ and satisfying: $0 \leq x \leq a$, $0 \leq y \leq a$, and $0 \leq z \leq a$. (See Exercise 7.)

Exercise 4 Use *Maple* to compute the area and center of mass of the first quadrant of the ellipse $(x/a)^2 + (y/b)^2 = 1$. Use this result to determine the area of the entire ellipse and make sure the specialization, $a = r$, and $b = r$, gives the correct area for a circle.

Exercise 5 Repeat the last problem for three dimensions, that is, compute the volume and center of mass of the portion of the ellipsoid $(x/a)^2 + (y/b)^2 + (z/c)^2 = 1$ that lies in the region $0 \leq x \leq a$, $0 \leq y \leq b$, and $0 \leq z \leq c$. Use this result to figure out the volume of the entire ellipsoid and make sure the specialization, $a = b = c = r$, gives the correct volume for a sphere.

Exercise 6 Repeat the last problem for *four* dimensions, that is, compute the hyper-volume and center of mass of the portion of the hyper-ellipsoid $(x/a)^2 + (y/b)^2 + (z/c)^2 + (w/d)^2 = 1$ that lies in the region $0 \leq x \leq a$, $0 \leq y \leq b$, $0 \leq z \leq c$, and $0 \leq w \leq d$. Use this result to figure out the hyper-volume of the entire hyper-ellipsoid and specialize the result to obtain the hyper-volume of the four-dimensional sphere of radius r.

AFTER THE LAB

Exercise 7 Show that the sum of your answer to Exercise 3 and the answer obtained in *Illustrative Example 1* is geometrically plausible.

Work in filling a tank

Consider the work done in filling a tank with a fluid. For definiteness, let's assume that the fluid is originally in a shallow lake at $z = 0$ and that the tank is elevated with its z-dimension between $z = a$ and $z = b$, see Figure 38. For realistic values of a and b,

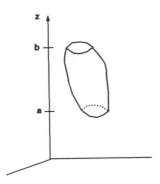

FIG. 38. Sketch of a body bounded between $z = a$ and $z = b$.

the constant force assumption is valid, that is, the force on a mass element Δm is just its weight $\Delta m \cdot g$. Thus the work done in raising this mass element of fluid is $\Delta W = z \cdot \Delta m \cdot g$, where z is the height to which this mass element is raised. Now $\Delta m = \rho \Delta x \Delta y \Delta z$, where ρ is the density of the fluid, so we obtain

$$W = \rho g \iiint_V z \, dx \, dy \, dz,$$

where the subscript V denotes integration over the volume of the tank. This is a familiar looking integral! We know that the z-component of the center of mass is given by

$$\bar{z} = \frac{\iiint_V z \, dx \, dy \, dz}{\iiint_V dx \, dy \, dz},$$

where the denominator is just the volume V of the tank. Thus $W = \rho g V \bar{z}$ or $\boxed{W = Mg\bar{z}.}$ Here, M denotes the total mass of the fluid filling the tank. The boxed result has two implications:

theoretical implication The work done in filling the tank is the same as the work in lifting a point particle of mass M a distance $\bar{z}$ against gravity, i.e., to the height of the z-component of the mass center.

practical implication If we know the height $\bar{z}$ for the tank in question, and we know the total mass M of the fluid that fills the tank, then computing the work is an easy calculation.

As an example of the second implication, consider the work done in filling a spherical tank of radius R whose lowest point is at $z = h$. Obviously $\bar{z}$ is the height of the center of the sphere, so $\bar{z} = h + R$. Since the mass is $M = \rho \cdot V = \rho \cdot 4\pi R^3/3$, we get $W = 4\pi \rho g R^3 (h+R)/3$ without doing any integrals!

Exercise 1 In using the integral to evaluate the work that we just calculated directly by using the boxed equation, it is better to measure z from the center of the sphere instead of from the ground. Show that in this case, the work integral becomes:

$$W = \rho g \int_{-R}^{R} \int_{ymin}^{ymax} \int_{xmin}^{xmax} (z + h + R) \, dx \, dy \, dz,$$

where the y and x-limits remain to be resolved.

Exercise 2 Use the method of cross sections to set up a one-dimensional integral for the work done in filling the spherical tank considered above (you are *not* required to evaluate this integral). Use the form of the work integral given in the previous problem.

Exercise 3 Repeat the previous problem using spherical coordinates to derive a three-dimensional integral for the work.

Exercise 4 Without doing any integrals, evaluate the work done in filling a vertical cylindrical tank resting on the ground with a fluid of mass density ρ. Assume that all the fluid is originally on the ground and that the cylinder has height H and radius R.

Exercise 5 Repeat the last problem for a horizontal cylinder.

The inverse square law

The $F = -mg$ force law is just an approximation to the inverse square law that is valid near the surface of the earth. It is natural to wonder if the mass center trick also works for the full fledged inverse square law, which reads

$$F = -\frac{GMm}{r^2}.$$

Below, we shall pursue the question: *Is the force between two bodies the same as the force between two point masses separated by a distance equal to the distance of the mass centers of the original bodies?* The inverse square law force between two mass particles with respective masses M and m and separated by a distance q has magnitude

$$F_{point} = -\frac{GMm}{q^2} \tag{33}$$

and acts along the line connecting the two particles.

IN THE LAB

Exercise 6 Use *Maple* to evaluate the integrals you set up in Exercises 1, 2 and 3 and thus check the simple evaluation that we obtained from the boxed equation.

Exercise 7 Use the boxed equation to help compute the work done in filling a tank in the first octant of the usual x-y-z coordinate system that is formed by the plane $x + y + z = 1$ and the coordinate planes. Use *Maple* to get the needed volume and z component of the mass center.

Exercise 8 In this Exercise, we'll investigate the force between a particle of mass m located at the point $(0,0,q)$ on the z-axis and a thin square plate of negligible thickness h and of density ρ filling the region $-a/2 \leq x \leq a/2$, $-a/2 \leq y \leq a/2$, $-h/2 \leq z \leq h/2$. By symmetry, the resultant force is wholly in the z-direction. This vertical force between the particle and a volume element of the box is given by

$$\Delta F = \frac{Gm\,\Delta M \cos \alpha}{R^2},$$

where $R = \sqrt{x^2 + y^2 + q^2}$, $\Delta M = \rho h\, \Delta x\, \Delta y$, and $\cos \alpha = q/R$. Note that if the thin plate were collapsed onto its mass center (the origin), then the force between the true particle and the fictitious one at the mass center of the plate would be given by Equation 33, where M is the total mass of the thin plate.

a) Draw a sketch and justify the formulas given above for R, ΔM, and $\cos \alpha$.

b) Assuming that the vertical variation in force is negligible over the small thickness h, show that
$$F = -Gmh\rho q \int_{-a/2}^{a/2}\int_{-a/2}^{a/2} \frac{1}{R^3}\, dx\, dy.$$

c) To make our formula for the force look more like the point force formula, eliminate the density ρ and show that the resulting expression is
$$F = -\frac{GMmq}{a^2} \int_{-a/2}^{a/2}\int_{-a/2}^{a/2} \frac{1}{R^3}\, dx\, dy.$$

d) Evaluate the integral in the previous part using *Maple*. Is $F = F_{point}$?

e) Take $a = 1$ and table the F/F_{point} ratio for $q = 2^k$, $k = 0, 1, \ldots 8$. Does $F \to F_{point}$ as q gets large? (Large q results are referred to as "far field" results in the scientific literature.)

f) We know what expanding an expression for *small* x means—get the Taylor series about $x = 0$. To get the series for *large* x in some sense means getting a series about $x = \infty$. And *Maple* actually understands this! For example (using the default for the number of terms):

```
series( a/sqrt(a^2 + q^2), q = infinity);
                3            5
               a            a           1
   a/q - 1/2  ----  + 3/8  ----  + 0(----)
                3            5           7
               q            q           q
```

Expand the F you obtained by integration for large q and thus show that one gets $F \approx F_{point}$ in the far field. (We will say more about this "After the Lab.")

Exercise 9 Repeat the previous problem for a thin circular disk of radius a. In evaluating the integral, use polar coordinates in the xy-plane.

AFTER THE LAB

What does it really mean to say that a distance q is "large" or is "small"? After all, suppose $q = 1\,\text{m}$. If we measure in microns, so that $q = 1,000,000\,\mu$, is q now suddenly a "large" distance? Or if we measure in kilometers, is $q = 0.001\,\text{km}$ suddenly a "small" distance? Of course not! In real problems, there are natural scales and other quantities are "large" or "small" *relative* to these scales. In the inverse square law problems we posed above, the quantity a provides the scale and large q really means that the dimensionless quantity q/a is large—or, equivalently, that a/q is small. Writing things in terms of $x = a/q$ reduces the mysterious expansions about ∞ to Taylor series about $x = 0$ and *that* we do understand!

Exercise 10 Justify the large q expansion given above for $a/\sqrt{a^2 + q^2}$. **Hint**: Factor a q out of the square root.

Exercise 11 Show that for the thin square plate, $F \to F_{point}$ as $x = a/q \to 0$. **Hint**: $\arctan x = x + O(x^3)$.

Illustrative Exercise We now treat a problem of great historical interest: *Find the force between a sphere and a particle that is external to the sphere.* From our experience with such problems in the *Force Applications in 3D* project, we expect that in the "far field," the force will be nearly the same as the force between the original particle and a fictitious particle located at the center of the sphere. Also from that experience, we would not expect this replacement to be exact. But Newton showed that it *is* exact! We will repeat his calculation using *Maple* to help us. Along the way, we'll discover that in this problem, *Maple* needs a little help from us to simplify the result correctly.

First of all, we pick the z-axis to be the line from the center of the sphere to the particle. Thus we can write the location of the particle as $(0, 0, q)$. Denote the radius of the sphere by a and its density by ρ. Since the particle is exterior to the sphere, we have $q > a$. With these choices, the geometry is as shown in Figure 39. By symmetry, the resultant force is

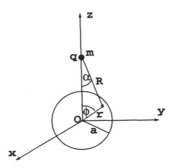

FIG. 39. Geometry of particle and sphere.

wholly in the z-direction. This vertical force between the particle and a volume element of the box located at the point with spherical coordinates, (r, ϕ, θ), is given by

$$\Delta F = \frac{Gm\Delta M \cos\alpha}{R^2}.$$

Here, by the law of cosines, $R^2 = q^2 + r^2 - 2qr\cos\phi$. Also, we have $\Delta M = \rho r^2 \sin\phi \Delta r \Delta\phi\Delta\theta$, and $\cos\alpha = (q - z)/R$, with $z = r\cos\phi$. Thus, the gravitational attraction is given by

$$F = -Gm\rho \int_0^a \int_0^{2\pi} \int_0^\pi \frac{q - r\cos\phi}{R^3} r^2 \sin\phi\, d\phi\, d\theta\, dr.$$

Finally, to facilitate comparison with the force law between two points, we eliminate the density by using its definition: $\rho = M/V = 3M/(4\pi a^3)$ to obtain

$$F = -\frac{3GMm}{4\pi a^3} \int_0^a \int_0^{2\pi} \int_0^\pi \frac{q - r\cos\phi}{R^3} r^2 \sin\phi\, d\phi\, d\theta\, dr.$$

Here is the evaluation of this integral using *Maple*:

```
R := sqrt( q^2 + r^2 - 2*q*r*cos(phi) ):
F := -3*G*M*m/(4*Pi*a^3) *
        int( int( int( (q - r*cos(phi))/R^3*r^2*sin(phi),
        phi = 0..Pi ), r = 0..a ), theta = 0..2*Pi );
                        F := 0
```

$F = 0$? The solution is wrong because *Maple* at one point took $\sqrt{(a-q)^2}$ and got $a - q$, having no way of knowing that $q > a$ (since the particle is *outside* the sphere). We were forced to intervene by the following trick. Before integrating, replace q by $a + eps$ and then compute F. Then replace eps by $q - a$ and simplify. (The shortcoming here is that *Maple* assumes $\sqrt{C^2} = C$, as we are also inclined to do). After this trick is done *Maple* returns the right answer:

$$F = -\frac{GMm}{q^2}.$$

The above 'trick' is implemented as follows. After defining R:

```
q := a + eps:
F := ...
q := 'q':
F := simplify( subs( eps = q-a, F ) );
```

IN THE LAB

Exercise 1 Consider a hollowed out sphere, that is, the solid portion is the region $a < r < b$. Find the force between the hollowed out sphere and a particle outside this body. (Note: you will need to use the above $q = a + eps$ idea to avoid an incorrect $F = 0$).

Exercise 2 Same as previous problem, but now the particle is located inside the hollow region.

Exercise 3 Consider a cylinder of radius a, height $2h$ and density ρ filling the region $0 \le x^2 + y^2 \le a^2$, $-h \le z \le h$. (Note: again you will need to use the above $q = a + eps$ idea.)

a) Derive a formula for the force F between this cylinder and a particle of mass m located at the point $(0, 0, q)$ on the z-axis. Write your result in terms of the mass M of the cylinder and evaluate using *Maple*.

b) Let F_{point} denote the force between the particle and a fictitious particle of mass M at the origin. Is $F = F_{point}$?

c) Take $a = 1$ and $h = 1$ and table the F/F_{point} ratio for $q = 2^k$, $k = 0, 1, \ldots 8$. Does $F \to F_{point}$ as q gets large (i.e., in the "far field")?

d) Expand the F you obtained by integration for large q and thus show that one gets $F \approx F_{point}$ in the far field.

Exercise 4 Consider a particle of mass m located at $(0, 0, q)$ on the z-axis and assume that a mass distribution of density ρ is spread out over the entire xy-plane. Show that the attraction between the plane and the point is given by

$$F = -G\rho mq \int_{-\infty}^{\infty}\int_{-\infty}^{\infty} \frac{1}{R^3}\, dx\, dy,$$

where $R = \sqrt{x^2 + y^2 + q^2}$. Evaluate the force using *Maple*.

AFTER THE LAB

Exercise 5 Do the integral in the *Illustrative Example* by hand. **Hint**: Use the law of cosines result for R^2 to replace the variable ϕ by the variable R.

Exercise 6 Does the resulting force in Exercise 4 bear any resemblance to the force between two particles? Comment on the physical meaning of the result.

BEFORE THE LAB

In this project we will seek solutions to differential equations, concentrating on the *nonhomgeneous* or *forced* equation. Typically, in applications, a system is acted upon by an outside force. For example, a mass in space will be attracted by the gravitational force of a larger body, resulting in a differential equation like $y''(t) = f$, or in the mass-spring problem an external force f is applied to the mass resulting in $my'' + Ky = f$.

We will first concentrate on determining approximate solutions in the form of a series, $y(x) = \sum_{k=0}^{\infty} a_k x^k$, valid for some interval about $x = 0$. In a computer or numerical implementation, one truncates this infinite series to a finite number of terms. Then the key questions are "How many terms should we take?" and, "Can we be confident of the resulting approximation?" Our approach is to expand the forcing term in a similar series, $f(x) = \sum_{k=0}^{\infty} b_k x^k$. Since we know the forcing term (in most applications), we assume that the b_k are likewise known. Then we substitute both the series for y and the series for f into the differential equation, equate like powers of x and solve for the a_k. If you did the project *Taylor Series—Advanced Usage*, then you may recall that *Maple* has an elegant means for handling this problem. However, we start with an approach more closely tied to what we'd do by hand.

Exercise 1 Solve the following differential equations with the initial conditions $y(0) = \alpha$, and $y'(0) = \beta$.

a) $y'' = 3x - 5x^3$. **Hint**: Just integrate twice.

b) $y'' = \sin x$.

c) $y'' = \sin x^2$. Since you can't integrate $\sin x^2$, seek a series for y. Get the first four non-zero terms of a series for $\sin x^2$ by replacing u by x^2 in the series for $\sin u$. Integrate the resulting approximate differential equation twice to get y.

Exercise 2 Now consider the forced mass-spring problem, $y'' + Cy = f$, where C is a constant and $f(x) = \sum_{k=0}^{\infty} b_k x^k$ with the b_k known. As discussed above, seek the solution in the form $y(x) = \sum_{k=0}^{\infty} a_k x^k$.

a) Derive the following *recurrence formula* for the a_k:

$$a_{k+2} = \frac{b_k - Ca_k}{(k+1)(k+2)}, \qquad k \geq 0.$$

b) For $f(x) = \sin x^2$, find the first five non-zero terms for y. Again use the initial conditions, $y(0) = \alpha$, and $y'(0) = \beta$.

IN THE LAB

Illustrative Exercise We find the first ten terms of the series solution to $y'' + y = 3\cos 2x$, with initial conditions, $y(0) = 2$, and $y'(0) = 0$. In the code given below, we first compute the b_k using the `coeff` command which extracts from P_n the coefficients of x^k. Then we compute the a_k from the recurrence formula of Exercise 2 and form the approximation y_n. In Figure 40, we compare y_n with the exact solution ($y = 3\cos x - \cos 2x$).

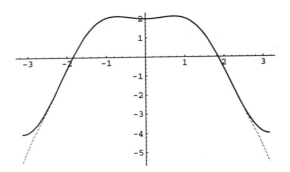

FIG. 40. Exact solution (dark) and approximate solution (light) for Illustrative Exercise.

```
f := x -> 3*cos(2*x):
kMax := 11:
Pn := taylor( f(x), x=0, kMax );
for k from 0 to kMax do
   b[k] := coeff( Pn, x, k)
od:
a[0] := 2:   a[1] := 0:
for k from 0 to kMax do
   a[k+2] := (b[k] - a[k]) / ((k+1)*(k+2))
od:
yn := sum( a[n]*x^n, n = 0..kMax+1);
```

$$Pn := 3 - 6x^2 + 2x^4 - \frac{4}{15}x^6 + \frac{2}{105}x^8 - \frac{4}{4725}x^{10} + O(x^{11})$$

$$yn := 2 + \frac{1}{2}x^2 - \frac{13}{24}x^4 + \frac{61}{720}x^6 - \frac{253}{40320}x^8 + \frac{1021}{3628800}x^{10} - \frac{4093}{479001600}x^{12}$$

```
plot( {yn, 3*cos(x) - cos(2*x)}, x = -Pi..Pi );
```

Exercise 3 Referring to the *Illustrative Exercise*:

a) Verify that $y = 3\cos x - \cos 2x$ is the exact solution to this initial value problem.

b) Note that the series for y_n has two more terms than the series for f. Why is that?

c) While the above approximate solution is visually adequate on the interval $[-\pi/2, \pi/2]$, it is not so for the interval $[-\pi, \pi]$. Take enough terms (kMax) so that you get good graphical agreement on $[-\pi, \pi]$.

Exercise 4 For the differential equation, $y'' + y = \ln(1+x)\cos 2x$, with initial conditions, $y(0) = 0 = y'(0)$, a solution in terms of elementary functions is no longer available. Hence a series, or other numerical approximation, is called for. Since we don't have an exact solution available to check against, we need other means to gain confidence that our approximate solution is good enough. Suppose we need an approximate solution that is graphically good on the interval $[0, \pi/4]$.

a) Compute enough coefficients b_k so that, graphically, your Taylor polynomial approximation of $f(x)$ is adequate. Record the maximum numerical error in your approximation.

b) Compute the coefficients a_k of the desired solution corresponding to your b_k. Then form the approximate solution, say, y_n. Graph y_n on $[0, \pi/4]$. Comment on your confidence in y_n at this point.

c) To check the accuracy of the above y_n, turn to the *variation of parameters* solution (discussed in any undergraduate text on Differential Equations):

$$y(x) = \int_0^x f(t) \sin(t - x) \, dt.$$

Verify that **int** fails to integrate this expression. Use **evalf(Int (...))** to get a graph of $y(x)$ on the interval $[0, \pi/4]$.

d) Compare y_n and $y(x)$, either graphically on $[0, \pi/4]$ or by building a table. Comment on the accuracy of y_n. In particular, how does its accuracy compare with that of the approximation to $f(x)$? Can you explain this?

A More Elegant Way in *Maple*. While it is important for you to understand the method for getting the recurrence formula in Exercise 2, the coefficients a_k can be computed in *Maple* without your finding the recurrence formula.

```
f := x -> 3*cos(2*x):
kMax := 13:
Pn := convert( taylor( f(x), x=0, kMax ), polynom );
                2      4        6         8          10            12
    Pn := 3 - 6 x  + 2 x  - 4/15 x  + 2/105 x  - 4/4725 x  + 4/155925 x

y := 'y':
diffequ := diff(y(x), x$2) + y(x) = Pn:

Order := 13:      #  this tells the next command to use 13 terms in the series
yseries := dsolve( {diffequ, y(0)=2, D(y)(0)= 0}, y(x), series);
yseries := y(x) =

            2    13   4    61  6    253   8    1021   10     4093    12       13
  2 + 1/2 x  - ---- x  + --- x  - ----- x  + ------- x  - --------- x  + O(x  )
              24        720      40320      3628800      479001600
```

Notice that the series agrees with that obtained earlier.

Exercise 5 Solve the problem in Exercise 4 using the above approach. Make sure it agrees with your earlier result.

Numerical solutions

In a course on differential equations, the topic of series solutions is usually developed in great detail. In particular, the methods introduced above are generalized to cover cases involving variable coefficients, singularities, nonlinear equations, etc. However, instead of pursuing series solutions further, we now give code for finding and displaying *numerical* solutions to differential equations using the `dsolve` command, with the `numeric` option, on the problem previously solved by series in the *Illustrative Example*. The theory behind such numerical solutions starts with *Euler's Method*, which is probably discussed in your text and certainly will be thoroughly covered and extended if you take a course in differential equations.

```
f := x -> 3*cos(2*x):
diffequ := diff(y(x), x$2) + y(x) = f(x):
fnum := dsolve( {diffequ, y(0)=2, D(y)(0)= 0}, y(x), numeric );
          fnum := proc(x) 'dsolve/numeric/result2'(x,1292292,[2]) end
fnum(Pi);
              {y(x) = -4.000000013, x = 3.141592654}
#  To plot the numeric solution, fnum:
with(plots):
odeplot( fnum, [x, y(x)], -Pi..Pi );
```

The resulting **plot** is shown in Figure 41.

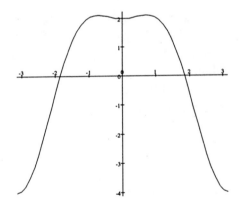

FIG. 41. Numerical solution with `dsolve` for Illustrative Exercise.

Exercise 6 Plot a numeric solution to the problem posed in Exercise 4.

AFTER THE LAB

Nothing this time.

BEFORE THE LAB

This project revisits the theme introduced in two earlier projects: *Work Along an Arc* and *Vectors and Work*. You now have enough mathematical tools to bring this subject to a logical conclusion. Until now the missing piece has been the notion of 'independence of path' for line integrals; that is, in going from point A to point B (in the plane or in R^3) does the path taken matter? If the application is arc length, for example, the obvious answer is 'yes'; and it comes as a bit of a surprise that in many line integral problems the answer is 'no.' The path independence issue is especially important in applications involving work, as one often wishes to minimize the amount of work, energy, etc. associated with a given task. In real life, the problem is usually compounded by other factors and trade offs are required. For example, if the required task is to move a load from point A to point B and the forces themselves are independent of the path taken, rarely does it *really* not matter about the path. Time is usually a factor, suggesting a straight line path. On the other hand, perhaps certain obstacles need to be honored, thus ruling out a straight line path. So the minimization problems suggested here, and elsewhere in your freshman-sophomore studies, are merely an introduction to those more realistic ones faced by modern science and industry. But the foundations are important, and we think that this foundation will be strengthened by this project.

Exercise 1 Suppose you are to power a boat from point $A = (0,0)$ to point $B = (2,1)$, and the primary consideration is the force of the wind which generally opposes you. You are to investigate the effect of taking different paths from A to B. Note: the wind forces considered have negative components, indicating that the wind opposes the direction of the path. Hence the Work done by the wind is negative. But one may also think of the Work as that required to counteract the wind, thus producing a positive value. We prefer the latter point of view.

a) Suppose the wind force is $F = (-a, -b)$, for a and b positive. Compute the work done along the the straight line path between A and B. Also compute the work done along a second path, of your choice. Does the path matter here? Why?

b) Now, due to the effect of the harbor you are entering, the wind force is $(-a, -ae^{-y})$. Again, compute the work along two paths as in part (a). Does the path matter here? Why?

c) Now let $F = (-a, -ae^{-y+cx})$ for $c > 0$. Compute the work along the straight line path and show that Work $= 2 + (1 - \exp(2c - 1))/(1 - 2c)$. (Note: for $c = 0$ you should agree with the Work in part (b)).

d) Now for a bit of a surprise. For the force of part (c), compute the work along the broken line path: up to (0,1) and over to (2,1). Explain why one does *not* have independence of path for this force. Comment on the fact that this path completely avoids the resistive part of the force represented by e^{cx}.

IN THE LAB

An important type of problem in several fields of application is: Can we find the 'optimal path' (e.g., to minimize the work)? You are to explore this issue in regard to the wind force in part (c) of the previous exercise, $F = (-a, -ae^{-y+cx})$. Along with the paths you used in Exercise 1, you are to consider some new paths. Consider first, a family of parabolas described for $0 \le t \le 1$, by:

$$x = 2t, \quad y = 2t(\beta - 2\alpha t) \quad \text{where } \alpha > 0.$$

Exercise 2 For the parabolic paths, and for the force $F = (-a, -ae^{-y+cx})$ with $c = 0.1$:

a) To go from (0,0) to (2,1), show that $\beta = (1 + 4\alpha)/2$.

b) Graphically show that for α 'large' the path swings 'north,' i.e. it goes above the line $y = 1$ before ending at $y = 1$. (This could be bad). What's the situation for $\alpha = 0$? Moreover, find α such that the path comes into the final destination horizontally. (This could be good. You may want to find α analytically).

c) Now compute the Work in going from (0,0) to (2,1) along various parabolic paths. What is the situation regarding the minimizing of Work, when there are no restrictions on the parabolic path? Is this surprising in any way?

d) Same problem as in part (c), except now you have the restriction that the path cannot, due to the coast line, swing 'north.' There *is* an optimal path in this case; what is it?

e) We have so far assumed $\alpha > 0$. What happens for $\alpha < 0$, both graphically and regarding the Work?

Exercise 3 Now consider a circuitous path from (0,0) to (1,2) described in *Maple* by:

```
x := t -> 2*t + sin(2*Pi*t):
y := t -> t^2 + 1 - cos(2*Pi*t):
```

where $0 \le t \le 1$.

a) Compute the arc length of this path.

b) Compute the Work on this path for the current force. How does it compare with the Work along the parabolic paths?

c) Compute the Work on this path for the force $F = (-a, -ae^{-y})$. Compare with the results of Exercise 1b), and comment.

AFTER THE LAB

Exercise 4 Reflect on the work of the previous exercises.

a) What role did the parameter c play in the amount of work required?

b) For $c > 0$, your experience should suggest that there *is* a parabolic optimal path in the case one cannot 'swing north.' What is it?

c) Is the path in (b) practical, say for navigating a boat?

d) What would you suggest as a practical path to (nearly) minimize the Work?

e) Recall that the work in Exercise 3(c) on the circuitous path was the same as that on the straight line path. How would you explain this strange result to a younger friend?

Notes

Notes

Notes

Notes

Notes

Notes

Notes

Notes

Notes

Notes